Bus Handbook

Mark Jameson

November 1995

British Bus Publishing

The Volvo effort into the Lynx programme is shown in LX2004, the first Lynx with a Volvo engine. This example went into service with Wrights of Wrexham who operated as The Wright Company. It is seen in Wrexham at the southern end of service 1 to Chester. The vehicle has since moved further north to Aintree Coachlines. *Malc McDonald*

Contents

The Leyland Lynx Bus Handbook

The Leyland Lynx Bus Handbook is the first in a series of volumes from British Bus Publishing detailing the production of specific types and models of impact on the British bus and coach scene. These are based on information kindly provided by the builders as production took place, and by operators and enthusiasts as vehicles have changed hands.

The writer and publisher would be glad to hear from readers should any information be available which corrects or enhances that given in this publication.

Principal Editors: Mark Jameson and Bill Potter

Acknowledgements:
We are grateful to David Donati, Ken Hargreaves, Steve Sanderson, the PSV Circle and the staff and managements of Leyland Bus and Volvo Bus for their co-operation in the compilation of this book. To keep up to date with the vehicle movements we recommend *Buses*, published monthly by Ian Allan Ltd, and the PSV Circle news sheets.

The front cover photo is by Tony Wilson

ISBN 1 897990 61 8
Published by British Bus Publishing Ltd
The Vyne, 16 St Margarets Drive, Wellington,
Telford, Shropshire, TF1 3PH
© British Bus Publishing, November 1995

LEYLAND LYNX - THE BACKGROUND

The Leyland Lynx was the final Leyland Bus single deck design to come into production. The model should be seen as the final chapter of a vehicle concept which started at the end of the 1960s rather than a market trend setter. The Lynx was also the result of a complex political history and it is necessary to go into the arrangements between operators and manufacturers to show its background in the market place.

After the end of World War 2, the Labour Government took much of British public transport into public ownership. Outside the London area the nationalised Tilling group (THC) formed one of the two major bus operations, the other being the British Electric Traction (BET) group which also had bus interests in several British Colonies. Many municipalities also ran their own services and local private operations were also widespread. In Scotland, the situation was slightly different in that the Scottish Omnibuses group operated alongside Alexanders and Caledonian MacBrayne.

In March 1968, BET sold its British interests to THC. Later that year the Labour Transport Minister, Barbara Castle, introduced the 1968 Transport Act that was initiated to found two new bodies to run buses in Britain. These were the Scottish Bus Group (SBG) which brought together Alexander's, Caledonian, David MacBrayne and Tilling, while for England and Wales the National Bus Company was born.

At this early stage, operators' boundaries were not re-drawn and were to continue in their traditional areas with service licences operated under the traffic commissioners remit until the next stage. This too was political and began the creation of Passenger Transport Authorities for the major conurbations. These brought about the disappearance of many of the larger municipal operators through a process that was mainly complete by the end of 1970. A side effect was that many NBC operations in the PTA regions were also transferred into their control. The effect of this action on the NBC was considerable. Midland Red, for example lost its most remunerative routes to the West Midlands PTE and North Western much of its work to SELNEC. At the same time London Transport lost its country operations to London Country Bus Services, a new company soon to become a subsidiary of the NBC.

Bus building was also going through major changes. The Jaguar group controlled Daimler and Guy, both of which had themselves earlier taken over other builders. Leyland Motors had taken over Standard-Triumph, the British Motor Corporation which controlled Austin and Morris, and AEC. In 1966 British Motor Holdings had been formed to oversee the activities of a merged Leyland and Jaguar group, and in 1968 the British Leyland Motor Corporation was formed. With this formation also came shares in Tilling's own manufacturers — Bristol and Eastern Coach Works. Following all these mergers a holding company, known as Bus Manufacturers (Holdings) Limited, was formed in 1970 to link the interests of BLMC and NBC.

Another facet of the 1968 Transport Act was the introduction of Bus Grants which were aimed at giving work to suppliers in a period of declining orders, improving vehicle standards and particularly at ensuring the safe implementation of pay as you board operation. Bus Grants were introduced at a 25% rate and later increased to 50% providing a major impetus to the operator to remove old vehicles by buying new vehicles.

At this time pay as you board was limited to single deck only and with the advent of extra funds for new buses, there was an influx of newly introduced rear-engined types with AEC, Leyland, Daimler and Bristol all competing for a share of this new market. The result was that the new designs had many problems. The Daimler Roadliner and AEC Swift (London called the heavier duty version of the latter chassis the Merlin) were notorious. The Leyland Panther and Panther Cub fared better while Bristol did best with the RE and this model was produced in several variants. The problems with these models were mainly technical, complex electrical systems and fluid couplings being the cause of many fires. When the engine was placed near the driver, problems could be smelled and heard before they became major. With the engine at the back this was not possible and breakdowns resulted.

Structural failures were also a problem and, as with much of the 'throw-away' 1960s, designed parts had a short service life. Education of traffic and engineering staffs became an urgent priority.

Another feature of this period, which has continued in a lesser way to the present day, was the issuing of grants to manufacturers to move to new green-field sites or re-development areas to assist with unemployment problems. Thus, by 1970, there was a completely new picture of grant-based industrial development, grants for new vehicles, a revised operating base structure and new controls by newly formed companies wishing to participate in the future.

Early in 1969 NBC and Leyland, through its BMH role, were trying to do several things, of which one was the improvement of the first generation rear-engined single deckers. Leyland had introduced its new 500-series engine at the 1968 Motor Show and were in the process of designing a completely new integral single deck bus to comply with all expected UK regulations of the time and also to meet the emerging EEC regulations. A hard task, still not complete in the 1990s, but a further problem was that many of its premises which BMH then occupied had served well for many years, but none of them were equipped for high volume standard specification output. At the time, Leyland

The first Leyland National prototype, P701 was built before the shape was determined. This example was used for much of the testing, including extreme low temperatures in Norway. It is seen during the tests. It later went to Spain for high temperature testing and was, unfortunately, chosen to be eventually crash tested at the MIRA proving ground.

manufactured bus chassis at Farington in Lancashire, AEC at Southall, Daimler at Coventry with bodywork output from Park Royal in north-west London, Roe at Leeds and Eastern Coach Works in Lowestoft. A very diverse manufacturing base!

The new concept bus would be built on assembly lines similar to car production methods and to a standard specification in a new purpose built factory. The seven prototypes of this vehicle, known as B7, were built at Farington but a new site in Cumbria, outside the steel town of Workington, was chosen for the building of a new assembly plant. Construction commenced on a site north east of Distington on the A595, at a point known as Lillyhall. Work on the new plant also included the funds to improve the nearby A66 trunk road from Penrith to Workington, a route which allowed access to west Cumbria without traffic affecting the English Lake District. Part of the agreement to have this new factory was that Bristol and Eastern Coach Works would close. This, in fact, did happen, not when originally intended but several years later. When the go-ahead for this work occurred shares were transferred from Leyland Vehicles and NBC to BMH.

The chassis of the Lynx is seen in this picture of one of the first examples photographed at Lillyhall prior to it being despatched to Alexander. During transit the body was strengthened by a transporter frame which is removed during bodybuilding.

The Leyland National, as the B7 became known, was announced in September 1970 and was a striking new integral single deck bus built to a rigidly controlled specification. Production commenced in a plant with a planned production of 22 units per week, and potential growth to 44. The new method of construction was revolutionary to the bus world and bought car production methods into this field.

The Farington built prototypes were numbered PP701 upward and were all subject to extensive testing. Workington's pre-production vehicles were numbered 00001 to 00010 and the production run commenced at 00101. NBC's first delivery was vehicle 00005 on the 13th March 1972 to the local subsidiary Cumberland Motor Services. It cost £8956. This price showed some thousand pounds saving on a comparable NBC Bristol RE/Eastern Coach Works single deck bus. The next pair for NBC went to London Country for the Stevenage Superbus operation in April 1972 and each cost £9089. The first production unit was to be a two-colour vehicle for SELNEC PTE, and thus visited the spray-booths twice. 00102 was all-over green for Crosville, and during the paint process passed 00101, to be the first vehicle production vehicle. Early production concentrated upon the 11.3m long two-door model. Other standard options were a single-door version, and similar models in the shorter 10.3m vehicle.

However even this expensively engineered vehicle was not without its teething problems and its standard, non-variable specification did not endear itself to a highly critical customer base. Rival manufacturing companies, who felt this new high volume production vehicle could be a threat to their own future, focused on its problems, presenting a 'down' on the successes of the National. Time, however, has shown that the vehicle had some real strengths, notably the end of the short-term, throw-away attitude.

Leyland also wanted to tackle the market outside the NBC. The fleet age profile in many of the PTAs was tipped in the balance of old vehicles and fleets needed rejuvenation. Leyland had always been successful in the export market and this area provided another opportunity to be tapped. For NBC, there was a planned 12 to 15 year life expectancy for heavy vehicles, and many of the vehicles inherited by London Country dated back to the late 1940s!

Hardly had the Leyland National gone into production when the Government lifted the ban on one-man operation of double deckers. This drastically reduced the potential market base for the National. Nevertheless, production continued, the NBC having placed its 1973 and 1974 programme orders, and the breed improved with lessons learnt.

However, overseas markets were not so ready to take in fully built up vehicles, or even Completely Knocked Down (CKD) vehicles for local assembly, as for political and financial reasons they preferred to use their local bodybuilder. So, in 1974, Leyland produced one 'chassis' based on National components and sent it to Eastern Coach Works for bodywork. In November 1974 this was displayed as the C27. Though destined to be a lone vehicle this unit was, in effect, the start of the Lynx. The market for this product was intended to be Australia, but it never went there. A further development, some six months later, was a chassis built at Lillyhall and known as the B21 and this product met the requirement for a local body to be fitted. Lillyhall later supplied the units to the Bristol plant, where they were built up. This version was for export only, but this included Ireland.

By the later 1970s production of many other Leyland products had been replaced by the National but there was one area in which it had not been able to compete, namely the very rural or lightly loaded operation. The Bristol LH, up to then, had covered this sector of the market with models which ranged from a short LHS (8m), standard LH (9 or 10m) to the LHL at 11m, all at both 2.3m and 2.5m widths. The LHL was mainly aimed at the coach market, but the LH and LHS were supplied for both bus and coach use. There was a regular demand for this simple vehicle, but a requirement by London Transport to have versions at 2.3m width and fitted with automatic gearboxes to replace ageing AEC Regal Mk IV (RF-type) buses brought the specification into conflict with the National.

The Leyland National was an integral vehicle with the running units and sub assemblies attached to the body in much the same way as a car is produced. To meet a demand for a chassis, which the Lynx was designed to meet, National 00432 was constructed with a perimeter frame. This was later bodied by Eastern Coach Works and known as C27.01. This 'chassis' version of the National is seen in Leyland on the testing ground.

Crosville also wanted to order this type of vehicle, but the NBC management put a hold on the requirement. BMH then introduced a 'B' variant of the National. This short version was a considerably lighter and less sophisticated vehicle without the overhead heating/demisting system, roof cladding, a down-rated 150bhp engine and black painted areas without brightwork. But even this could not be built at 2.3m width, and the LHS variant was to continue for several more years to satisfy this market.

Several operators built up large fleets of this lighter and cheaper B-type, notably London Country. However the 510 engine, standard to the National, despite being new, was small and fast revving and had failed to gain popularity with operators.

By the late 1970s, customers were asking for the Leyland 0.680 unit to be incorporated. In addition, 'green' issues were taking hold and emission control limits were having to be met. Leyland's answer was shown at the 1978 Motor Show when the prototype National 2 with the 0680 engine was displayed to the public. In November 1979 the model was launched at the Scottish Motor Show. The market place at the time was still being affected by national politics, the Conservative party was back in power and, from April 1980, the Bus Grant was reduced from 50% to 40%. Thereafter it was reduced by 10% each year, being phased out in March 1984.

In 1981 the NBC wrote off the loss of the Workington factory from its accounts. In that year it sold its 50% share of its holdings in BMH to Leyland Vehicles giving the latter total ownership of the Leyland National assembly plant and production facilities at Bristol, Eastern Coach Works and Roe. The NBC was under much political pressure not to incur vehicle manufacturing losses so the company instituted changes to improve the situation. In March 1981, the passenger vehicle division of Leyland Vehicles became Leyland Bus. The old Guy factory building export vehicles was closed and production moved to Farington. By now AEC and Daimler were history, but the lengthy delays in new vehicle deliveries experienced in the 1979/80 period had been caught up by 1981. The double deck market was also part of a major review of types. Until 1980 NBC ordered around 500 Leyland Nationals a year. Policy changes and reduced finance resulted in only around 50 being ordered for 1982 (the 1981 programme was cut back as a result of late 1979/80 programmed deliveries).

Dennis had also entered the market with a competitive vehicle. Known as the Falcon, it offered the Gardner 6HLXB engine that operators had been asking Leyland to incorporate for years. The effect was that while Leyland had not appeared to take any notice of its customers, it suddenly found its market share vanishing. Quickly, Leyland introduced the Gardner engine to the Mark 2 National but, with the damage already done, this did little to revive its market share.

By now Leyland Bus was researching into other markets to make use of surplus capacity at the National factory. Some of this was used to build the Titan from 1981 after the Park Royal plant in London was closed. Overall, London Transport took 1125 of these, but only small numbers were sold elsewhere. The Bristol plant at Brislington was latterly building the Olympian, a separate chassis to carry an independent body based on the Titan, and when Brislington closed in 1990, production of this, too, moved to Lillyhall.

During this period secret talks were going on with British Rail concerning the construction of a lightweight railcar, based on many of the National's body components. Small numbers of prototypes was built for demonstration in Britain and overseas, and a BR standard MK1 coach was rebodied with a Leyland National body for evaluation and this has, incidentally, now been sold for presevation. After lengthy delays and many changes of specification, Leyland eventually gained orders for 20 two-car units followed by 94 units plus an additional 35 of an exended 23-metre version.

It was also clear, from the early 1980s, that the coach market had become more important. At Roe the Royal Tiger Doyen coach had been introduced, available both as a complete vehicle and as an underframe that was licenced to Plaxton or Van Hool for bodying. The NBC management was still providing considerable input to the introduction and technical aspects of this coach. At Roe there was a skilled coachbuilding work-force, unlike Lillyhall which provided a cheaper semi-skilled team. Production there was more of a unit assembly plant but, even with this skill, there were many problems and very delayed deliveries. Even when the vehicles were delivered, problems plagued the vehicle. The problems were not all of Roe's making, but caused by a lack of accuracy in the Leyland parts and drawings. Roe was also producing good quality double-deck bus bodies at that time.

In 1984 Leyland Bus decided, as it had done with Bristol, to close the Roe plant and transfer work, in this case the Doyen, to Lillyhall. The Roe management, supported by West Yorkshire Enterprise Board, bought out the factory now known as Optare.

By late 1983 it was clear that the day of the National was fast drawing to a close with only low volume orders drifting in, whereas the plant relied on high volume production for economic viability. The re-election of the Conservative party in 1983 heralded further changes in the industry for, as part of its manifesto, it had planned the sale of the bus industry to the private sector.

By the middle of 1984 Lillyhall was producing only small numbers of National 2s, Royal Tigers (as both under-frames and complete coaches), the Olympian chassis, the last of the order from London Transport for Titans, and a handful of Class 141 railcars. Hardly the mass production line intended in 1972.

B60-03 was first built as a dual doored layout, though later modified and used for promotional work. When the new Leyland Truck plant was opened in May 1986 the vehicle was parked in the display area along with a National 2 demonstrator and Royal Tiger development vehicles. *Mark Jameson*

Leyland Bus still had the Farington factory as well and the experimental shop where various 'B' and 'C' designs came and went. One, coded the B60, was a project started in 1982 which was planned to replace the National with a vehicle available both as an under-frame (in the same manner as the Olympian in relation to the Titan), or as a complete vehicle. The intention was to appeal to overseas markets where local bodybuilders were protected from overseas competition, at a cost to home body assembly. As usual there were conflicting interests with the sale of such vehicles.

However, in 1984, the NBC had effectively stopped being interested in normal sized single deck buses, and had commenced investment in minibuses derived from vans. This policy change was largely led by Harry Blundred's development at Exeter where 16-seaters were able to reach locations that bigger buses could not reach, and where the perceived 'friendliness' of a small, compact vehicle gave feelings of greater security for its passengers. At this time too, the taxi was seen as a major threat to NBC revenue.

Opposite: Two further Lynx prototypes are seen here. In the livery of the Australian operator MTT is seen B60-02, an example with Bolton's bodywork which includes air conditioning pods. Following much success in exporting Nationals to the Australian continent, this early Lynx prototype was hoped to trigger many sales of the type. Also seeking prospective orders was B60-04, the model used by National Bus to evaluate the type. It was allocated to Ribble who used it from its Blackburn depot mostly on the 225 service between Clitheroe and Bolton. Photographed in Darwen, heading south it shows the step arrangement of the original Lynx. *Volvo Bus/Bill Potter*

The Leyland Lynx

Then the Bill that led to the 1985 Transport Act was published. The NBC was to be broken up and sold off and with it ended any hopes Leyland Bus may have had in persuading the NBC to take large numbers of their new B60. Only 44 Nationals were sold in 1984 and 57 in 1985. In November 1985 the last National was delivered to Halton. This carried vehicle number 07835. Another problem for Leyland Bus was the decision of Technical Services Department of NBC. NBC had always had standard specifications for all suppliers to follow, but were now looking for the suppliers to compete for building entirely to NBC specification. An idea too late on the political scene, and ironic at a time when London Transport had abandoned all further plans of building their own standard types. During the investigation by the NBC of the B60, the complete vehicle was offered to the NBC at £53,000 for an integral build vehicle adapted to NBC specification, but none was ever ordered, despite much hard work by Leyland Sales force.

Development of the B60, now named Lynx, continued in 1984 with the first vehicle delivered with an Alexander body for Citybus in Belfast. This was the first of six and the type illustrated on the rear cover. The Leyland management had given special permission for the Alexander body though it was intended that the entire home market would be complete vehicles bodied at Lillyhall. Six pre-production vehicles were built at Lillyhall and numbered B60.01 to B60.06. One went to Australia, one worked with Ribble Motor Services from Blackburn and the rest were experimental units. The first production unit was a chassis built by September 1985.

Orders were very slow to arrive. The balance of the Ulster Bus order was completed and trial orders were received from Greater Manchester and West Midlands PTEs. 1986 saw only 25 Lynx registrations. It was also a year for redundancies in the industry and the sale of parts of National Bus meant that finance was concentrated upon buy-outs rather than re-equipment. Deregulation took place in October 1986. The initial result was duplication of operators on routes, again reducing what revenue was available and constantly changing priorities left the bus industry insecure.

In July 1986 there was an announcement by the Government that they were prepared to sell Leyland Bus to its management team for £11.7m, and included a 33% stake in the Leyland parts operation. The buy-out was finally completed on 13th January 1987 and included the Workington and Farington factories, service centres at Bristol, Chorley, Glasgow and Nottingham and Leyland Bus Hong Kong together with rights in the spare parts DAF/United Bus consortium.

An immediate affect was the introduction of Cummins engines to specification options. In June 1987 the Conservatives were re-elected to government, showing that the political position was unlikely to be changed.

The Caldaire Group, which operates in Yorkshire, was a major customer for the Lynx, and the prime customer in its early days. Seen here is 327, G327NUM which is LX1633. Interestingly, the Caldaire Group have taken the replacement for the Lynx, the Volvo B10B, in large numbers also.
David Cole

In 1987 some 60 new Lynx were licensed, an improvement on the earlier year, but hardly profitable. Early in 1988, the Caldaire Group placed an order for 50 Lynx buses. Another Lynx built during this period was a demonstrator which received a special Alexander body. It was first shown at the UITP conference in Singapore in 1988 and was in the livery of Singapore Bus Services to whom it later went on extended loan. Even so, the new directors of Leyland Bus were not happy with the group's performance. The proposed single European market was a cause of concern and, with such a perceived threat, the management began looking for a new partner or association.

Volvo had been competing directly with Leyland Bus in the coaching field with the B10M, and sales of Tiger coaches were under threat as a result. Then, after a mere 15 months of ownership by management, Leyland Bus became part of the Volvo corporation in a deal that surprised even the press. The deal was reputed to be for around £23m.

Following the sale to Volvo, and in an effort to improve viability, two models were soon dropped - the Royal Tiger Doyen, and the Leyland Lion. The latter was a double-deck chassis built by DAB in Denmark with many British parts. This also competed with the Volvo B10M Citybus that was modified at Irvine into a double-deck chassis for the UK market. About the same time as the Doyen was stopped, production

Standing inside the Lillyhall compound ready for delivery are four examples for Midland Red West. It was normal practice to set the destination indicators to the last three digits of the chassis numbers while the vehicles were at the factory. The building in the background was the test centre while the main factory is to the right of the picture. *Mark Jameson*

Showing the rear end arrangement of the Lynx is LX1291, the first purchased by the London Borough of Hillingdon and operated by London Buses. The vehicle has now joined the main CentreWest fleet as LX1, F101GRM, and is seen in 607 Express livery. *Colin Lloyd*

The early demonstrator, much travelled and photographed, was LX1425, registered F74DCW and which has now joined the Red & White fleet. Its original livery was similar to that used by West Midlands Travel and was photographed in that scheme in Blackpool while on loan to them in 1989.
Paul Wigan

also ceased of the competitive, semi-integral, C10M built by Ramseier & Jenser in Switzerland, which was sold in Britain in limited numbers while the Doyen never achieved exports.

Volvo also tried to improve the export market of the Lynx. One underframe was shipped out to Portugal, and was bodied there by Camo, but it did not lead to further orders. It subsequently returned to Leyland to join the experimental fleet at Farington and is still in that role. Volvo also found that some of the large build orders reported by the Leyland prior to their take over had been stock-build rather than revenue earning customer orders, a fact that the new owners had not contemplated in the take-over calculations.

During 1988 the Eastern Coach Works plant at Lowestoft, after completing an order for 250 Olympian bodies for London Transport, was closed. The jigs were then transferred to Lillyhall where some of the lessons learned in building the body were incorporated into the re-sited build line. Rationalisation of sales forces which still had the Leyland team and also Volvo's own sales operation in Warwick took place in 1989. A new sales operation was set up based in Warwick under the name of VL Bus and Coach UK Limited. This operation looked after sales in the UK and in Ireland.

LX1108 was used for demonstration in 1988 before joining Stagecoach's Cumberland operation in 1989 along with three new examples. Seen in Aldershot while on loan to Alder Valley E709MFV shows off the scheme which clearly shows its purpose. *Ralph Stephens*

As Volvo gradually became more involved with Lillyhall production, they tried to improve Lynx popularity by incorporating Volvo ideas though they were up against constraints of technology caused by earlier lack of investment in premises equipment. Late in 1988 MCW, Leyland/Volvo's chief rival in the UK and with British exports, was put up for sale by its owners, the Laird Group. MCW made the Metrocab and trains as well its bus range.

Orders for the Lynx began to pick up slightly during 1989, first with sales utilising vehicles from stock and then by new build batches. Then came an order for 23 vehicles from Badgerline for its Bristol City Line operation and, the only sizable order for the Lynx, for 150 vehicles for West Midlands Travel. On the down side, Volvo's Swedish management had prevented sales of the Lynx at the Barcelona show - it later emerged that they were working on a replacement, Volvo based, vehicle. However, despite these orders, the interest in new vehicles was still declining, partly as a result of a further nationwide recession in industry.

Lillyhall was still having over-capacity problems even with all body production concentrated at the plant and, by the start of 1990, the position was even worse. On the credit side, Volvo had commenced building their popular B10M chassis in the UK though these were assembled from kits imported from Sweden, the first being haded over to Park's of Hamilton. Sales of Farington-produced Tiger and Swift models continued to tumble and Volvo were faced with the decision that

Farington would have to close because of the recession. At that time the Laird Group had not found a buyer for MCW and other vehicle builders had closed including Duple, which had been taken over by Plaxton in 1991. Lynx production away from the West Midlands Travel order was steady with further orders from amongst others, the Badgerline group, but not at the rate of the National that preceded it.

In early 1990 Volvo management were having a re-think about the future of the Lynx and in June 1990 the first Lynx II appeared. The new DiPTAC design features were incorporated and, of necessity to take the Volvo engine, the vehicle gained a snout which increased the overall length slightly. None of this prevented further redundancies at Lillyhall and, despite development work to make new models, sales of Lynx and other full-size buses were still falling overall.

On 6th December 1991 Volvo announced that the British operation would be restructured. Volvo already had a manufacturing plant at Irvine in Scotland, with much land to develop further if necessary. Lillyhall would be closed with the progressive loss of nearly 400 jobs. The Lynx, which had continued with stock-build examples, would be terminated. The plan to build the new B6 at Workington was dropped in preference to the longer-established Irvine operation. Olympian chassis production would also transfer there, its fourth location (Leyland and Bristol the other two) but Olympian body production would cease.

The final production Lynx was delivered to Halton in August 1992 though the last new Lynx registrations were licensed from stock in September to Brewer's and Alder Valley.

Lillyhall received a short reprieve when a major order for Olympian chassis was received from Singapore Bus Services and a major fire in Glasgow resulted in a large order for the type from Strathclyde Buses. The final Strathclyde Buses' chassis eventually left the plant in July 1993, thus ending a major chapter in British Bus building.

Brighton was one of several municipal operators to take the model. LX1131 is pictured as Brighton 47, E447FWV during the 1988 summer.
Mark Jameson

In total some 1058 Lynx were built, not a prolific type and not too common to see, unless you now live in the West Midlands or the West Riding of Yorkshire. Now the future is in the hands of its operators. As far as is known only one Lynx has been lost in an accident and the other vehicles whose current locations are unknown are from the prototype status. Where Lynx have appeared on the second-hand market they appear to have been snapped up quickly, which is a good sign.

In today's market, with a demand for low floor access and DiPTAC specifications, the Lynx and National still have places - the purpose built vehicle with low entrance like these vehicles are in demand. The writer is not convinced by the engineering competence of suspension lowering mechanisms, particularly as vehicles get older.

It is now nearly ten years since the Lynx first appeared and the whole market place has turned again. Minibus growth has stabalised and the dominant size of single decker appears to be in 33-43 seat range. Volvo has the B6 for this range and the low-floor version, B6LE. In the larger single deck range the B10B rear-engined chassis and the recently introduced B10L low floor and B10LA low floor articulated version. As demand for new vehicles increases only history will tell if Volvo's decision to drop the Lynx was the right one.

Opposite: **A reproduction of the technical specification leaflet issued by Leyland to potential customers. This provides further details of the vehicles component specifications and options available.**

Lynx City Bus

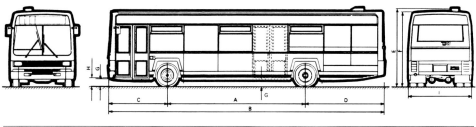

NOMINAL CHASSIS DIMENSIONS

MODEL	A	B	C	D	E	F	G	H	I
LYNX	5630mm	11 180mm	2370mm	3180mm	3129mm	3000mm	305mm	340mm	2500mm

Entrance door step heights: Rear half 'G', Front half 'H'; Exit door step height: 'G'
Floor heights laden. Front — 500mm; Rear — 950mm

APPROXIMATE VEHICLE WEIGHTS

MODEL	CERTIFIED UNLADEN			MAX DESIGN WEIGHT			Outer Swept Circle
	Front axle	Rear axle	Total	Front axle	Rear axle	Total	
LYNX Bus	2960kg	5953kg	8913kg	6000kg	10 500kg	16 500kg	21.55m
LYNX Dual Purpose	2990kg	6063kg	9053kg	6000kg	10 500kg	16 500kg	21.55m

PERFORMANCE GUIDE

MODEL	GVW (kg)	ENGINE	GEARBOX	AXLE RATIO	TYRES	MAX GEARED SPEED (mph)	MAX CLIMB	MAX RESTART
LYNX	14 000	6HLXCT-205	4HP500	4.656	275/70R 22.5	43.3	2.5	3.1
				4.19		47.9	2.8	3.5
				3.7		54.6	3.3	4.0
LYNX	14 000	L10-B210H	4HP500	4.659	275/70R 22.5	43.3	2.3	2.9
				4.19		47.9	2.4	3.2
				3.7		54.6	3.0	3.8
LYNX	14 000	6HLXCT-234	5HP500 Dual	4.659	275/70R 22.5	45.7 (55.0)	2.1	2.6
				4.19		50.5 (60.8)	2.4	2.9
LYNX	14 000	L10-B220H	5HP500 Dual	4.659	275/70R 22.5	49.2 (59.3)	2.3	2.9
				4.19		54.4 (65.3)	2.6	3.2

( ) denotes 5th gear mode

LEYLAND LYNX - PRODUCTION

Looking back on the Lynx now, former Leyland employees may be surprised and delighted to learn how devoted some operators of the model have become, how the model has faired, how few enter the second-hand market and where they have, how quickly they have been snapped up. The vast majority of changes in the owners list are through company changes which have accelerated since the 1986 Transport Act. Judging by the low sales of the first years, it was not always so.

Leyland Bus Limited gave the first technical information to the principal customer, the National Bus Company, in May 1986. With much acclaim they provided the following initial specification:-

"The Lynx 11.18 metre overall length standard model represents the latest generation of single deck vehicles and is offered with a range of optional engines and transmission units together with alternative door and seating layouts. The vehicle is of integral construction with the body consisting of a welded framework of rectangular hollow section tube fully integrated with the welded open section rolling underframe. Stressed upper cove panels are incorporated in strategic positions to give added structural integrity. Aluminium alloy exterior panels which are easily replaced for accident repair and direct glazing are standard design features.

In the standard form the vehicle includes seats for 51 passengers and 22 standees can be carried on normal stage carriage operation. Alternative door and seating configurations are available.

The Leyland Lynx is supplied with the Leyland TL11 turbo charged engine rated at 210 bhp at 1850 rev/min together with four speed automatic transmission incorporating an integral retarder as standard equipment. Alternative engines and transmission units are available if required, i.e. Cummins L10H and Gardner 6HLXCT engines and ZF4HP500 gearbox. Full air suspension, power assisted steering and dual air brakes are also standard features."

The letter went on to supply drawings of the standard layout, and full technical specification.

Three interesting aspects stand out here; first there was a standard build vehicle with 51 seats which became the most numerous type, and secondly the offer of alternative engines. Both Gardner and Leyland had been installed into the prototypes but none carried the Cummins code. Most notable was the reference to an integral vehicle as far as the home market was concerned. The promotional leaflet issued at this time gave further information, and this is reproduced on page 23.

Meanwhile, in the export arena, Leyland bus announced two models of Lynx, the LX563TL11FR and the LX563TL11FL, the difference being right or left hand drive models respectively. Here the designed overall

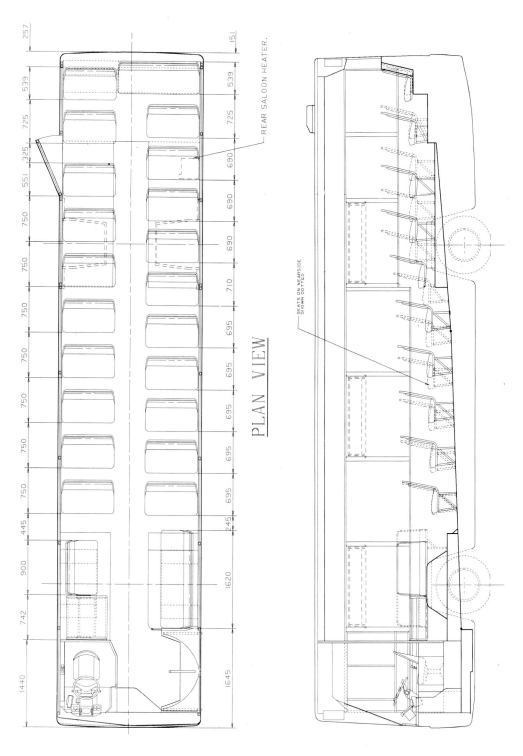

PLAN VIEW

REAR SALOON HEATER.

SEATS ON NEARSIDE
SHOWN DOTTED

Following de-regulation, Shearings diversified into local service work. As with its coach operation, quality new vehicles were purchased to reduce the cost of maintenance. Two batches of Lynx were delivered along with Leyland Tigers and, later, Volvo B10M buses. Seen near Ironbridge while heading for Telford town centre is LX1296, F48ENF. While the main operation continues with Timeline, the Lynx have been sold, this example going to Maidstone & District. *Bill Potter*

length was 11 metres with the same wheelbase as the complete vehicle of 5.63 metres. The remainder of the specification also followed that for the UK market. During delivery of the chassis a rigid frame was added above the chassis for extra strength.

Back in the UK, two vehicles were available for sale at this time, both of which were pre-production models already painted poppy red. These were offered to the NBC during 1968. One, B60.04 which had been photographed with dummy plates C805LNX, was not taken up. The other, B60.05, was taken into NBC ownership and allocated to Ribble for evaluation work. They numbered it 901 and placed it in service at Blackburn. A year earlier Ribble had also taken the last of an order for ten Leyland Tigers with Duple bus bodywork, the other nine meeting a requirement for Midland Red North.

Production commenced late in 1985 and stopped again in 1986 after fifteen units had been completed. These included six evaluation buses for WMPTE and six Alexander-bodied chassis for Ireland. The three remaining were all development units which were extensively tested with two dismantled without sale. The main production got under way during the summer of 1986 with an evaluation vehicle for SBG, run by Kelvin Scottish and four for Greater Manchester. Other interesting deliveries were for Colchester and keen National operator Halton.

Vehicle LX1023 was the first with a Cummins engine, the vehicle becoming a demonstrator once development work was complete.

There were two principal floor options with the Lynx, the ramp and a step version which was introduced later. The stepped unit retained a flat floor up to the rear axle then a single step to another level over the engine. The ramp option eliminated this step by a gradual incline which levelled out over the rear axle. From LX1016 onwards this variant was added as a suffix to the model code. The remainder of the code was broken down as follows and, while the plan consisted of five vehicle lengths, only the standard 112 model became reality.

LX	Indicated the Lynx product
104	10m overall length (never used)
107	10.7m overall length (never used)
109	10.9m overall length (never used)
110	11m overall length (never used)
112	11.18 overall length
120	12m overall length (never used)
TL11	Leyland TL11 engine
6LXB	Gardner 6LXB engine
6LXCT	Gardner 6LXCT engine
L10	Cummins L10 engine
F	Fluid coupling Leyland Hydracyclic gearbox
T	Torque coupling Leyland Hydracyclic gearbox
V	Voith D851 gearbox
Z	ZF HP500 gearbox
L	Left hand drive (never used)
R	Right hand drive
1	Number of doors when assembled
R	Ramped floor
S	Stepped floor

A typical code would be LX112TL11FR1R.

The Lynx chassis is seen here in an official picture. The engine fitted is a Volvo THD102 KF and believed to be LX1211 photographed before the vehicle was sent to Portugal to gain its Camo body. The vehicle is back in Britain, and is used for further development work on new Volvo products. The Volvo B10B and B10L owe much to the Lynx in their development..

LX1033 was the only production Lynx built to the dual door specification, though produced as a speculative build. After a period in store it was modified and sold to the Caldaire Group who placed it into the Sheffield & District fleet. The vehicle is seen here as 254, E254TUB. *D Akrigg*

The first dual-doored example was LX1033 which was built for dealer stock in white livery. This turned out to be the only example of the original Lynx in this form, the only dual-doored examples in use are a dozen Lynx IIs with Lothian. LX1033 had the centre door removed in the autumn and was sold to Sheffield & District in December 1987.

Production continued during 1987 and 1988 with several deliveries to interesting independents such as D Coaches in south Wales, Safeguard, Shearings, Fishwick, AA Motors, Moor-Dale and Jubilee of Stevenage. London Buslines introduced the model to London for the tendered service 81 and Merthyr Tydfil took fifteen for a major modernisation programme.

By Christmas 1987 the only former NBC fleet to have taken the type was the Caldaire group. The now private NBC fleets started to look at full-size single-deck bus replacements. West Yorkshire RCC ordered a pair and Western National six. Municipal fleets took more interest in 1988 with further examples for Halton and Colchester, and they were introduced into Nottingham, Chesterfield and Brighton.

Tendered operations dictate the size of vehicle and in some cases required new vehicles to be used. Pan Atlas won the 112 service for London Regional Transport and took new Lynx. Also during 1988, Eastern National introduced the type with an order of 30. Repeat orders came from Shearings and there were more for the West Riding fleet. The turn of the year saw a focus on the north east with an order for six from Busways and twelve from Go-Ahead, its first order for large buses since it was privatised in a management buy-out in 1987. As with most operators, the Lynx was used to improve service quality and Go-Ahead placed the type on its Sunderland services.

By the end of 1988 the plant had in production some 230 vehicles though several were still not allocated to customers. The 1988 transfer to Volvo made little change, the model continued to carry the Leyland name though an additional engine option, the Volvo THD0102 was added. The first vehicle with this power unit did not appear until a chassis for the Portuguese market was bodied by Camo of Oporto - a bodybuilder connected with Volvo which has also built buses on B10M, B10R and 1960s Leyland Royal Tiger.

In November 1988 a new list of 'rationalised' model codes was introduced. These followed a new system for the whole product range and changed the earlier types. While the memorandum suggested the codes should be adopted as soon as possible, the first recorded Lynx with a new code was chassis LX1211 for Portugal, with the first complete bus LX1237, first of a trio for Citybus of Southampton with a stepped floor followed by a pair with high-back seating and a ramped floor. Official records also suggest that, unlike the changes to National codes, part orders were completed with the whole batch displaying the existing number.

The new description took the following aspects in order: model; number of axles; left or right hand drive; wheelbase or overall length; engine and gearbox. The intention was that where a model required to show additional information in the code, e.g. 8" brakes on the Tiger or the stepped/ramped floor on the Lynx, these additional fields would be added at the end of the code after an oblique symbol. We are not aware if this aspect appeared on the vehicles. The codes for the Lynx were:

LX	Lynx
2	2-axle
L	Left hand drive
R	Right hand drive
11	11.18m overall length (bodied)
55	5.5m wheelbase underframe
56	5.63m wheelbase underframe
61	6.11m wheelbase underframe
C15	Cummins L10-210 (157kW)
C16	Cummins L10-220 (164kW)
G15	Gardner 6HLXCT-205 (153kW)
G17	Gardner 6HLXCT-234 (175kW)
V18	Volvo THD0102 (180kW)
Z4	4HP500 gearbox
Z5	5HP500 gearbox
R	Ramped floor
S	Stepped floor

A typical code would be LX2R11C15Z4R.

Two Lynx joined the Isle of Man Road Services fleet in 1990 following on from several batches of Nationals. These started a new numbering series and seen in Douglas is 1, BMN401T. *Daniel Hill*

Displaying the engine switches and radiator filler arrangement is this rear view of demonstrator E709MFV taken at the 1988 Brighton Coach Rally. *Mark Jameson*

The first Lynx to enter service with the Stagecoach group - then the largest independent bus operator in the UK - were four for the Cumberland operation in June. Three were new Cummins-engined examples, the fourth was LX1108 which had served some time as a demonstrator. They were joined in 1991 by B60.05 after its period with Ribble had ended. The vehicle by then had undergone a major overhaul and engine replacement to bring it in line with the new models. These were the only Lynx to be ordered by Stagecoach, other Volvo products with Alexander bodywork, notable the B6 and B10M completing the remainder of the order.

The Caldaire Group, with the West Riding operation the focus, placed a further order for thirty-five in the summer. Those destined for United were delivered the following January while those for Yorkshire followed in the spring. Deliveries throughout 1989 showed a marked increase in production, the first signs that the minibus revolution had peaked. Significant new customers for the product were Cardiff, Cleveland Transit, Preston and a further order from West Midlands Travel, the operator of the original vehicles some three years earlier. Having left a gap of some three years since then, deliveries re-started and this fleet grew to be the largest Lynx operation with a total of 257 vehicles, almost 24.3% of all Lynx production.

The other major order for 1989 was from the Badgerline Group with deliveries to City Line at Bristol during 1989 and Midland Red West and City Line in 1990, with further examples for the Badgerline operation itself. These orders were the mainstay of production for the remainder of 1989 and 1990.

Stagecoach took three new Lynx and the then current demonstrator into the Cumberland fleet in 1989. These were joined in 1991 by B60-04 after its transfer from Ribble and with the specification changed to match the new examples. LX1312 is one of the three new vehicles all of which are now based at Penrith after a period at Barrow where they were operating when this photograph was taken. *Bill Potter*

Lynx II started a new chassis production series at LX2001, though the development had taken place using series 1 chassis. Following on from many earlier Lynx, City of Nottingham took one early example of the Lynx II with a further pair afterwards. LX2119 is seen heading for Beeston as Nottinghams 759, J759DAU. *Mike Fowler*

LYNX II

The latter part of 1990 saw much work behind the scenes in the development of a Mark II Lynx, which was unveiled on 10th December 1990. By now the Lynx was the market leader in the single-deck field, though this was still in small quantities in comparison with the days of regulated services and National Bus. Indeed, despite the low volume, the Lynx had achieved 100% of the 1987 single-deck bus registrations.

The new product showed much influence from Volvo, though the Leyland name was to continue. There were changes to both the exterior and the interior, the extended front being the most noticeable. Many DiPTAC features were incorporated, including an improved entrance step, green stanchions, low mounted palm-operated bell pushes, *bus stopping* sign and new non-slip flooring. On the mechanical side the modified specification included a new gearbox and revised front suspension which gave a better ride. This was achieved by specifying softer air springs combined with new shock absorbers. Two engines were offered in the revised model, the almost standard Cummins L10 turbo-charged diesel and the Volvo THD102KF.

The basic structure of the Lynx II remained, again based on a strong spaceframe that supported aluminium panels. A single entrance step height of 305mm was introduced, replacing the previous split arrangement. The rear-seat squab was now a 5-seat, deleting the notch back effect of the original design and the rear wheel arch moulding, now in one piece, reduced the intrusion into foot space for certain seats. The driver's compartment had been updated, with a reduced-height cab door to allow for the variety of modern ticketing equipment and a new instrument panel derived from the Tiger coach was used. The front header was also remoulded into a flat, full-width hinged panel that gave improved access to the destination equipment.

Two evaluation vehicles were built after an earlier stock model had been used to check the aesthetics of the design. One of these was taken to the Hallerad proving ground in Sweden, part of the development centre for Volvo. The first production fully-built vehicle with the Volvo engine went to The Wright Company of Wrexham in January 1991, the first Cummins examples to Halton.

The Caldaire group again proved an enthusiastic operator, though neither of the other two major operators of the original version, Badgerline Group and West Midlands Travel took the new product. During 1991 interesting deliveries included repeat orders from Nottingham and Cardiff, with Fishwick and Whitelaw's showing the small independents were still interested. An order for a batch of dual-doored examples for Lothian Regional Transport was delivered in March 1991. These are the only examples of the dual-doored layout to enter service in that form. As production came to an end early in 1992, the remaining stock models were sold, Halton, a keen supporter of the products from Lillyhall, took the last four produced. The final pair to be sold went to Alder Valley just prior to its sale to Stagecoach South, though this pair moved into The Bee Line operation in January 1993.

The successor to the Lynx is the Volvo B10R, which continues many of the features from the Lynx.

After a period on Demonstration to MTT, B60.02 moved to Lever of Quearbeyer where it gained an orange and blue scheme with Action names. The vehicle is seen in that scheme. Note that at that time it was still known as the B60 in certain areas.

PROTOTYPE VEHICLES

B60.01	LX5636LXBFR	Leyland Bus			Alexander	B37D	06/84
		City Bus	3006	HXI3006	Alexander	B37F	12/85
		Ulsterbus	3006	HXI3006	Alexander	B37F	?
		Stevensons	256	HXI3006	Alexander	B37F	02/95
B60.02	LX	Leyland Bus		6YY320	Boltons	B39D	01/85
		Lever, Quearbeyer		ZIB671	Boltons	B37F	8/87
		Deanes		ZIB671	Boltons	B37F	7/90
B60.03	LX112TL11FR2	Leyland Bus			Leyland	B37F	01/85
B60.04	LX112TL11FR1	Leyland Bus		(C605LNX)	Leyland	B51F	01/85
		Index mark used for photocall only					
B60.05	LX112TL11FR1	Leyland Bus		C544RAO	Leyland	B51F	01/85
		Ribble	901	C544RAO	Leyland	B51F	7/86
		Cumberland	255	C544RAO	Leyland	B51F	1/91
B60.06	LX563	Leyland Bus			Chassis only		01/85
		Leyland Bus			Leyland	B0D	

Initially chassis only, a body shell was fitted to B60.05 for a short period before the vehicle was dismantled.

EXPORTED KITS

LXK001	LX563L10ZR	Hill, Woollongong	Pressed Metal	B45D	09/89
LXK002	LX563L10ZR	Hill, Woollongong	Pressed Metal	B45D	09/89

These kits were ordered for UTA, Sydney.

PRODUCTION VEHICLES

LX1001	LX563TL11VR	Leyland Bus			Chassis only			09/85
		Stored at Lillyhall then dismantled						*06/87*
LX1002	LX5636LXBFR	City Bus	3007	HXI3007	Alexander	B43F		00/86
		Ulsterbus	3007	HXI3007	Alexander	B43F		?
		Stevensons	257	HXI3007	Alexander	B37F		02/92
LX1003	LX5636LXBFR	City Bus	3008	HXI3008	Alexander	B43F		00/86
		Ulsterbus	3008	HXI3008	Alexander	B43F		?
		Stevensons	258	HXI3008	Alexander	B37F		02/92
LX1004	LX1126LXCTFR1	Leyland Bus		C920FMP	Leyland	B51F		00/86
		West Riding	252	C920FMP	Leyland	B51F		05/87
		Sheffield & District	252	C920FMP	Leyland	B51F		12/87
		West Riding	252	C920FMP	Leyland	B51F		10/89
LX1005	LX5636LXBFR	City Bus	3009	HXI3009	Alexander	B43F		02/86
		Ulsterbus	3009	HXI3009	Alexander	B43F		09/89
		Stevensons	259	HXI3009	Alexander	B53F		01/92
LX1006	LX5636TL11FR	City Bus	3010	HXI3010	Alexander	B43F		02/86
		Ulsterbus	3010	HXI3010	Alexander	B43F		09/89
		Stevensons	260	HXI3010	Alexander	B49F		02/92
LX1007	LX5636TL11FR	Ulsterbus	3011	HXI3011	Alexander	B43F		09/85
		City Bus	3011	HXI3011	Alexander	B43F		09/89
		Stevensons	261	HXI3011	Alexander	B53F		01/92
LX1008	LX563TL11FR	Ulsterbus	3012	HXI3012	Alexander	B43F		02/86
		City Bus	3012	HXI3012	Alexander	B43F		09/89
		Stevensons	262	HXI3012	Alexander	B53F		02/92
LX1009	LX1126LXCTFR1	West Midlands PTE	1061	C61HOM	Leyland	B48F		02/86
LX1010	LX1126LXCTFR1	West Midlands PTE	1062	C62HOM	Leyland	B48F		02/86
LX1011	LX1126LXCTFR1	West Midlands PTE	1063	C63HOM	Leyland	B48F		02/86
LX1012	LX1126LXCTFR1	West Midlands PTE	1064	C64HOM	Leyland	B48F		03/86
LX1013	LX1126LXCTFR1	West Midlands PTE	1065	C65HOM	Leyland	B48F		03/86
LX1014	LX1126LXCTFR1	West Midlands PTE	1066	C66HOM	Leyland	B48F		03/86
LX1015	LX563TL11FL	Leyland Bus			Chassis only			03/86

Development chassis only, dismantled in 1987 with parts used to build LX1211.
After chassis LX1015 production was halted for several weeks to evaluate production to date.

LX1016	LX5636LXCTFR1	Dodds of Troon		D330LSD	Leyland	B52F		08/86
LX1017	LX5636LXCTFR1	Dodds of Troon		D340LSD	Leyland	B52F		08/86
LX1018	LX5636LXCTFR1	Kelvin Scottish	1401	D401MHS	Leyland	B51F		09/86
		Stevensons	60	D401MHS	Leyland	B51F		06/91
LX1019	LX563TL11FR1	Halton	34	D711SKB	Leyland	B51F		09/86
LX1020	LX563TL11FR1	Mackie, Alloa		D122FLS	Leyland	B49F		09/86
LX1021	LX5636LXCTZR1	Greater Manchester	501	D501LNA	Leyland	B48F		10/86
		GM North	501	D501LNA	Leyland	B48F		12/93
		GM North	1401	D501LNA	Leyland	B48F		01/95
LX1022	LX112TL11FR1	Colchester	31	D31RWC	Leyland	B49F		10/86
		Crosville Wales	SLL31	D31RWC	Leyland	B49F		02/94
LX1023	LX112L10ZR1	Leyland Bus		D634BBV	Leyland	B51F		06/87
		Mackie, Alloa		D634BBV	Leyland	B51F		04/88
LX1024	LX112LXCTZR1	Greater Manchester	502	D502LNA	Leyland	B48F		10/86
		GM North	502	D502LNA	Leyland	B48F		12/93
		GM North	1402	D502LNA	Leyland	B48F		02/95
LX1025	LX112LXCTZR1	Greater Manchester	503	D503LNA	Leyland	B48F		10/86
		GM North	503	D503LNA	Leyland	B48F		12/93
		GM North	1403	D503LNA	Leyland	B48F		02/95
LX1026	LX112TL11FR1	Colchester	32	D32RWC	Leyland	B49F		10/86
		Crosville Wales	SLL32	D32RWC	Leyland	B49F		02/94

LX1040 was new to Boro'line of Maidstone but moved to Kentish Bus in 1992. Numbered 418, D156HML it is seen in Penge shortly after the new livery was applied. *Malc McDonald*

While Ribble evaluated the product on behalf of National Bus, Kelvin took the Scottish Bus Group example where it was numbered 1401, D401MHS. It was mostly found on Dumbarton town services and is seen working service D2. The vehicle has since returned to England and is now with Stevensons. *Bill Potter*

LX1027	LX112LXCTZR1	Greater Manchester	504	D504LNA	Leyland		B48F	10/86
		GM North	504	D504LNA	Leyland		B48F	12/93
		GM North	1404	D504LNA	Leyland		B48F	02/95
LX1028	LX112TL11FR1	AA Motors (Youngs)		D573LSJ	Leyland		B51F	10/86
		AA Motors (Dodds)		D573LSJ	Leyland		B49F	04/92
LX1029	LX112TL11FR1	Safeguard		D165HML	Leyland		B49F	07/87
LX1030	LX112TL11FR1	Southern Motorways		D204FBK	Leyland		B50F	12/86
		Solent Blueline		D204FBK	Leyland		B50F	10/87
		West Riding	253	D204FBK	Leyland		B50F	02/88
		Sheffield & District	253	D204FBK	Leyland		B50F	02/88
		West Riding	253	D204FBK	Leyland		B51F	10/89
LX1031	LX112TL11FR1	John Fishwick & Sons	25	D25VCW	Leyland		B47F	11/86
LX1032	LX112TL11FR1	John Fishwick & Sons	30	D30VCW	Leyland		B47F	11/86
LX1033	LX1126LXCTFR2	Leyland Bus	u/r		Leyland		B45D	01/87
		Sheffield & District	254	E254TUB	Leyland		B51F	12/87
		West Riding	254	E254TUB	Leyland		B51F	10/89
LX1034	LX112TL11ZR1	D Coaches, Morriston		D32MWN	Leyland		B51F	01/87
		Whitelaw, Stonehouse		D32MWN	Leyland		B51F	01/91
		Bleasdale, Liverpool		D32MWN	Leyland		B51F	02/93
LX1035	LX112TL11ZR1S	London Buslines	751	D751DLO	Leyland		B49F	05/87
		The Bee Line	751	D751DLO	Leyland		B49F	10/95
LX1036	LX112TL11ZR1S	Jubilee, Stevenage	13	E970MNK	Leyland		B49F	12/87
		London Country NE		E970NMK	Leyland		B49F	01/89
		Sovereign, Stevenage	LX970	E970NMK	Leyland		B49F	01/89
		Luton & District	406	E970NMK	Leyland		B49F	05/90
		The Shires	406	E970NMK	Leyland		B49F	04/95
LX1037	LX112TL11ZR1S	Boro'line, Maidstone	227	D155HML	Leyland		B49F	05/87
		Kentish Bus	417	D155HML	Leyland		B49F	07/92
LX1038	LX112TL11ZR1S	London Buslines	752	D752DLO	Leyland		B49F	05/87
		The Bee Line	752	D752DLO	Leyland		B49F	10/95
LX1039	LX112TL11ZR1S	London Buslines	753	D753DLO	Leyland		B49F	06/87
		The Bee Line	753	D753DLO	Leyland		B49F	10/95
LX1040	LX112TL11ZR1S	Boro'line, Maidstone	228	D156HML	Leyland		B49F	05/87
		Kentish Bus	418	D156HML	Leyland		B49F	07/92
LX1041	LX112TL11ZR1S	Boro'line, Maidstone	229	D157HML	Leyland		B49F	05/87
		Kentish Bus	419	D157HML	Leyland		B49F	07/92
LX1042	LX112TL11ZR1S	London Buslines	754	D754DLO	Leyland		B49F	05/87
		The Bee Line	754	D754DLO	Leyland		B49F	10/95
LX1043	LX112TL11ZR1S	London Buslines	755	D755DLO	Leyland		B49F	06/87
		The Bee Line	755	D755DLO	Leyland		B49F	10/95
LX1044	LX112TL11ZR1S	London Buslines	756	D756DLO	Leyland		B49F	06/87
		The Bee Line	756	D756DLO	Leyland		B49F	10/95
LX1045	LX112TL11ZR1R	Merthyr Tydfil	101	D101NDW	Leyland		B51F	03/87
		Grey-Green	801	D101NDW	Leyland		B51F	12/89
		Grey-Green	801	D101NDW	Leyland		B49F	04/95
LX1046	LX112TL11ZR1R	Merthyr Tydfil	102	D102NDW	Leyland		B51F	03/87
		Grey-Green	802	D102NDW	Leyland		B51F	12/89
		Grey-Green	802	D102NDW	Leyland		B49F	04/95
LX1047	LX112TL11ZR1R	Merthyr Tydfil	103	D103NDW	Leyland		B51F	03/87
		Metrobus, Orpington		D103NDW	Leyland		B51F	11/89
LX1048	LX112TL11ZR1R	Merthyr Tydfil	104	D104NDW	Leyland		B51F	03/87
		Metrobus, Orpington		D104NDW	Leyland		B51F	12/89
LX1049	LX112TL11ZR1R	Merthyr Tydfil	105	D105NDW	Leyland		B51F	03/87
		Hillingdon LB	LX9	D105NDW	Leyland		B51F	12/89
		Hillingdon LB	LX9	809DYE	Leyland		B51F	05/90
		London Buses	LX9	809DYE	Leyland		B51F	09/91
		Centre West	LX9	809DYE	Leyland		B51F	09/94

LX1050	LX112TL11ZR1R	Merthyr Tydfil	106	D106NDW	Leyland	B51F	03/87
		Hillingdon LB	LX10	D106NDW	Leyland	B51F	12/89
		Hillingdon LB	LX10	810DYE	Leyland	B51F	03/90
		London Buses	LX10	810DYE	Leyland	B51F	09/91
		Centre West	LX10	810DYE	Leyland	B51F	09/94
LX1051	LX112TL11ZR1S	John Fishwick & Sons	32	D32YCW	Leyland	B47F	04/87
LX1052	LX112L10ZR1	Leyland Bus		D603ACW	Leyland	B51F	04/87
		Jubilee, Stevenage	11	D603ACW	Leyland	B51F	10/87
		London Country NE		D603ACW	Leyland	B51F	01/89
		Sovereign, Stevenage	LX603	D603ACW	Leyland	B51F	01/89
		Luton & District	405	D603ACW	Leyland	B51F	05/90
		The Shires	405	D603ACW	Leyland	B51F	04/95
LX1053	LX112TL11ZR1S	John Fishwick & Sons	33	D33YCW	Leyland	B47F	05/87
LX1054	LX112TL11ZR1R	Merthyr Tydfil	107	D107NDW	Leyland	B51F	06/87
		Mackie, Alloa		D107NDW	Leyland	B51F	11/89
LX1055	LX112TL11ZR1R	Merthyr Tydfil	108	D108NDW	Leyland	B51F	06/87
		Grey-Green	803	D108NDW	Leyland	B51F	12/89
		Grey-Green	803	D108NDW	Leyland	B49F	04/95
LX1056	LX112TL11ZR1R	Merthyr Tydfil	109	D109NDW	Leyland	B51F	06/87
		Cynon Valley	50	D109NDW	Leyland	B51F	11/89
		Red & White	394	D109NDW	Leyland	B51F	08/92
LX1057	LX112TL11ZR1R	Merthyr Tydfil	110	D110NDW	Leyland	B51F	06/87
		Metrobus, Orpington		D110NDW	Leyland	B51F	11/89
LX1058	LX112TL11ZR1R	Merthyr Tydfil	111	D111NDW	Leyland	B51F	06/87
		Hillingdon LB	LX11	D111NDW	Leyland	B51F	12/89
		Hillingdon LB	LX11	811DYE	Leyland	B51F	03/90
		London Buses	LX11	811DYE	Leyland	B51F	09/91
		Centre West	LX11	811DYE	Leyland	B51F	09/94
LX1059	LX112TL11ZR1R	Merthyr Tydfil	112	D112NDW	Leyland	B51F	06/87
		Mackie,Alloa		D112NDW	Leyland	B51F	11/89
LX1060	LX112TL11ZR1R	Merthyr Tydfil	113	E113RBO	Leyland	B51F	08/87
		Cynon Valley	51	E113RBO	Leyland	B51F	11/89
		Red & White	395	E113RBO	Leyland	B51F	08/92
LX1061	LX112L10ZR1	Moor-Dale	66	E677DCU	Leyland	B51F	12/87
		Northumbria	66	E677DCU	Leyland	B51F	11/94
LX1062	LX112L10ZR1	Moor-Dale	67	E678DCU	Leyland	B51F	12/87
		Northumbria	67	E678DCU	Leyland	B51F	11/94
LX1063	LX112TL11ZR1R	Jubilee, Stevenage	12	E371YRO	Leyland	B51F	11/87
		London Country NE		E371YRO	Leyland	B51F	01/89
		Sovereign,Stevenage		E371YRO	Leyland	B51F	01/89
		County Bus		E371YRO	Leyland	B51F	02/89
		County Bus	LX371	E371YRO	Leyland	B51F	04/89
		Sovereign, Stevenage		E371YRO	Leyland	B51F	10/90
LX1064	LX112TL11ZR1R	Merthyr Tydfil	114	E114RBO	Leyland	B51F	12/87
		Cynon Valley	52	E114RBO	Leyland	B51F	10/89
		Red & White	396	E114RBO	Leyland	B51F	08/92
LX1065	LX112TL11ZR1R	Merthyr Tydfil	115	E115SDW	Leyland	B51F	01/88
		Cynon Valley	53	E115SDW	Leyland	B51F	11/89
		Red & White	397	E115SDW	Leyland	B51F	08/92
LX1066	LX112TL11ZR1S	Grey-Green	885	E885KYW	Leyland	B47F	10/87
		Grey-Green	885	E885KYW	Leyland	B49F	04/95
LX1067	LX112TL11ZR1S	Grey-Green	886	E886KYW	Leyland	B47F	10/87
LX1068	LX112TL11ZR1S	Grey-Green	887	E887KYW	Leyland	B47F	10/87

Opposite, top: **LX1143 is seen here in the Atlas Bus livery used for the LRT tendered operation of the 112 service. The vehicle moved to Luton and District in 1994 where it gained a red and cream livery and has recently joined The Shires where a further scheme is used.**

Opposite Bottom: **Another LRT tendered operation brought six Lynx into the London Buslines operation. Seen in Buslines yellow livery, the vehicles have recently gained full livery for The Bee Line operation in Berkshire.**

When new to Grey Green in 1987 the Lynx wore a white, brown and orange livery as shown by 885, E885KYW photographed while working service 379 at Chingford Station. The paint process allowed the lower panels to be affixed after paint, and thus it was common for liveries to follow the wheelarch contour. *Philip Hambling*

LX1069	LX1126LXCTZR1R	West Yorkshire RCC	1201	E324SWY	Leyland	B49F	12/87
		York City & District	1201	E324SWY	Leyland	B49F	01/89
		Harrogate & District	324	E324SWY	Leyland	B49F	07/90
		Keighley & District	213	E324SWY	Leyland	B49F	04/95
LX1070	LX1126LXCTZR1S	Grey-Green	888	E888KYW	Leyland	B47F	10/87
LX1071	LX1126LXCTZR1S	Grey-Green	889	E889KYW	Leyland	B47F	10/87
LX1072	LX1126LXCTZR1R	West Yorkshire RCC	1202	E325SWY	Leyland	B49F	12/87
		York City & District	1202	E325SWY	Leyland	B49F	09/88
		Harrogate & District	325	E325SWY	Leyland	B49F	07/90
		Keighley & District	214	E325SWY	Leyland	B49F	05/95
LX1073	LX1126LXCTZR1S	Grey-Green	890	E890KYW	Leyland	B47F	10/87
LX1074	LX112L10ZR1R	Shearings	21	E21UNE	Leyland	B49F	10/87
		Yorkshire Traction	281	E21UNE	Leyland	B51F	08/91
LX1075	LX112L10ZR1R	Shearings	22	E22UNE	Leyland	B49F	10/87
		Yorkshire Traction	282	E22UNE	Leyland	B51F	08/91
LX1076	LX112TL11ZR1R	Jubilee, Stevenage	12	E840EUT	Leyland	B51F	11/87
		London Country NE		E840EUT	Leyland	B51F	01/89
		Sovereign, Stevenage	840	E840EUT	Leyland	B51F	01/89
		County Bus		E840EUT	Leyland	B51F	02/89
		County Bus	LX840	E840EUT	Leyland	B51F	04/90
		Sovereign,Stevenage		E840EUT	Leyland	B51F	10/90
LX1077	LX112TL11ZR1R	Halton	43	E641VFY	Leyland	B51F	11/87
LX1078	LX112TL11ZR1R	Shearings	23	E23UNE	Leyland	B49F	11/87
		Yorkshire Traction	283	E23UNE	Leyland	B49F	08/91
LX1079	LX112TL11ZR1R	Shearings	24	E24UNE	Leyland	B49F	11/87
		Yorkshire Traction	284	E24UNE	Leyland	B49F	08/91

New to Merthyr Tydfil in June 1987, LX1056 moved to Cynon Valley in 1989 and has subsequently joined the Red & White fleet. It is seen in Aberdare in September 1990 displaying the Cynon Valley livery of white, green and orange. *Malc McDonald*

LX1080	LX112TL11ZR1R	Stevenson's	72	E72KBF	Leyland	B51F	01/88
LX1081	LX112TL11FR1S	Safeguard, Guildford		E51MMT	Leyland	B51F	11/87
LX1082	LX112TL11FR1S	Western National	200	E200BOD	Leyland	B51F	02/88
LX1083	LX112TL11FR1R	Chesterfield	62	E62WDT	Leyland	B49F	12/87
		Rhondda	500	E62WDT	Leyland	B49F	07/94
LX1084	LX112TL11FR1R	Chesterfield	63	E63WDT	Leyland	B49F	12/87
		Birmingham Omnibus		E63WDT	Leyland	B49F	07/94
		Rhondda	502	E63WDT	Leyland	B49F	08/94
LX1085	LX112TL11FR1R	Chesterfield	64	E64WDT	Leyland	B49F	12/87
		Birmingham Omnibus		E64WDT	Leyland	B49F	07/94
		Rhondda	503	E64WDT	Leyland	B49F	08/94
LX1086	LX112TL11FR1R	Chesterfield	65	E65WDT	Leyland	B49F	12/87
		Rhondda	501	E65WDT	Leyland	B49F	07/94
LX1087	LX1126LXCTZR1S	Sheffield & District	255	E255TUB	Leyland	B49F	12/87
		West Riding	255	E255TUB	Leyland	B49F	10/89
LX1088	LX1126LXCTZR1S	Sheffield & District	256	E256TUB	Leyland	B49F	12/87
		West Riding	256	E256TUB	Leyland	B49F	10/89
LX1089	LX1126LXCTZR1S	Sheffield & District	257	E257TUB	Leyland	B49F	12/87
		West Riding	257	E257TUB	Leyland	B49F	10/89
LX1090	LX1126LXCTZR1S	Sheffield & District	258	E258TUB	Leyland	B49F	01/88
		West Riding	258	E258TUB	Leyland	B49F	10/89
LX1091	LX1126LXCTZR1S	Sheffield & District	259	E259TUB	Leyland	B49F	01/88
		West Riding	259	E259TUB	Leyland	B49F	10/89
LX1092	LX1126LXCTZR1S	Sheffield & District	260	E260TUB	Leyland	B49F	01/88
		West Riding	260	E260TUB	Leyland	B49F	10/89
LX1093	LX1126LXCTZR1S	Sheffield & District	261	E261TUB	Leyland	B49F	01/88
		West Riding	261	E261TUB	Leyland	B49F	10/89

Several independents have chosen the Lynx. Seen in Sudbury is LX1104, E87KGV which has been with H C Chambers & Son since new in March 1988. This example features the Cummins engine which, at that time, was less popular than either the Gardner or Leyland alternatives. *Colin Lloyd*

LX1094	LX1126LXCTZR1S	Sheffield & District	262	E262TUB	Leyland	B49F	01/88
		West Riding	262	E262TUB	Leyland	B49F	10/89
LX1095	LX112L10ZR1R	Colchester	33	E33EVW	Leyland	B49F	01/88
		Crosville Wales	SLC33	E33EVW	Leyland	B49F	05/94
LX1096	LX112L10ZR1R	Colchester	34	E34EVW	Leyland	B49F	01/88
		Crosville Wales	SLC34	E34EVW	Leyland	B49F	05/94
LX1097	LX112L10ZR1R	Colchester	35	E35EVW	Leyland	B49F	02/88
		Crosville Wales	SLC35	E35EVW	Leyland	B49F	05/94
LX1098	LX112L10ZR1R	Colchester	36	E36EVW	Leyland	B49F	02/88
		Crosville Wales	SLC36	E36EVW	Leyland	B49F	05/94
LX1099	LX1126LXCTZR1S	Sheffield & District	263	E263TUB	Leyland	B49F	02/88
		West Riding	263	E263TUB	Leyland	B49F	10/89
LX1100	LX1126LXCTZR1S	Sheffield & District	264	E264TUB	Leyland	B49F	02/88
		West Riding	264	E264TUB	Leyland	B49F	10/89
LX1101	LX112TL11ZR1R	Chesterfield	60	E60WDT	Leyland	DP45F	02/88
		East Midland	60	E60WDT	Leyland	B49F	08/95
LX1102	LX112TL11ZR1R	Chesterfield	61	E61WDT	Leyland	DP45F	02/88
		East Midland	61	E61WDT	Leyland	B49F	08/95
LX1103	LX112L10ZR1R	Colchester	37	E37EVW	Leyland	B49F	02/88
		Crosville Wales	SLC37	E37EVW	Leyland	B49F	05/94
LX1104	LX112L10ZR1R	H C Chambers & Son		E87KGV	Leyland	B52F	03/88
LX1105	LX112TL11ZR1S	Pan Atlas	AB57	E965PME	Leyland	B49F	07/88
		Atlas Bus	LL65	E965PME	Leyland	B49F	08/94
		Yellow Bus, Stoke Mandeville		E965PME	Leyland	B49F	11/94
		The Shires	422	E965PME	Leyland	B49F	07/95

Opposite: Two vehicles from the same batch new to Chesterfield in 1988. In the upper view is LX1083, E62WDT in the yellow and blue Chesterfield livery. The lower picture shows the Rhondda scheme applied to LX1086, E65WDT as Rhondda 501. All bar two of the Chesterfield Lynx were sold in recent years. The pair that remained have now joined the East Midland fleet following that operators acquisition of Chesterfield during 1995. *Tony Wilson*

A loyal customer for the National was Ayrshire coast operator AA Buses who also took three Lynx in 1986 and a further pair in 1989. Seen leaving Ayr for Kilwinning is F85XCS, chassis LX1331.
Malc McDonald

LX1106	LX112TL11ZR1S	Pan Atlas	AB52	E966PME	Leyland	B49F	07/88
		Atlas Bus	LL66	E966PME	Leyland	B49F	08/94
		Yellow Bus, Stoke M'deville	966	E966PME	Leyland	B49F	11/94
		The Shires	423	E966PME	Leyland	B49F	07/95
LX1107	LX112L10ZR1R	Halton	44	E49WEM	Leyland	B49F	03/88
LX1108	LX112L10ZR1R	Leyland Bus		E709MFV	Leyland	B49F	04/88
		Cumberland	254	E709MFV	Leyland	B49F	07/89
LX1109	LX112TL11ZR1R	Western National	201	E201BOD	Leyland	B51F	03/88
LX1110	LX112TL11ZR1R	Western National	202	E202BOD	Leyland	B51F	03/88
LX1111	LX112TL11ZR1R	Western National	203	E203BOD	Leyland	B51F	03/88
LX1112	LX112TL11ZR1R	Western National	204	E204BOD	Leyland	B51F	03/88
LX1113	LX112TL11ZR1R	Western National	205	E205BOD	Leyland	B51F	03/88
LX1114	LX112TL11ZR1R	Jubilee, Stevenage	15	E420EBH	Leyland	B49F	03/88
		London Country NE	LX420	E420EBH	Leyland	B49F	01/89
		Sovereign, Stevenage	LX420	E420EBH	Leyland	B49F	01/89
		County Bus, Stevenage	LX420	E420EBH	Leyland	B49F	01/89
		Sovereign, Stevenage	LX420	E420EBH	Leyland	B49F	10/90
LX1115	LX112TL11ZR1R	Jubilee, Stevenage		F359JVS	Leyland	B49F	08/88
		London Country NE	LX359	F359JVS	Leyland	B49F	01/89
		Sovereign, Stevenage	LX359	F359JVS	Leyland	B49F	01/89
		County Bus, Stevenage	LX359	F359JVS	Leyland	B49F	02/89
		Sovereign, Stevenage	359	F359JVS	Leyland	B49F	10/90
LX1116	LX112TL11ZR1R	Jubilee, Stevenage		F358JVS	Leyland	B49F	08/88
		London Country NE	LX359	F358JVS	Leyland	B49F	01/89
		Sovereign, Stevenage	359	F358JVS	Leyland	B49F	01/89
LX1117	LX112TL11ZR1R	City of Nottingham	725	E725BVO	Leyland	B51F	03/88
LX1118	LX112TL11ZR1R	City of Nottingham	726	E726BVO	Leyland	B51F	03/88
		City of Nottingham	726	E726BVO	Leyland	B49F	11/92

LX1119	LX112TL11ZR1R	City of Nottingham	727	E727BVO	Leyland	B51F	03/88
LX1120	LX112TL11ZR1R	City of Nottingham	728	E728BVO	Leyland	B51F	03/88
		City of Nottingham	728	E728BVO	Leyland	B49F	??
LX1121	LX112TL11ZR1R	City of Nottingham	729	E729BVO	Leyland	B51F	04/88
		City of Nottingham	729	E729BVO	Leyland	B48F	??
LX1122	LX112TL11ZR1R	City of Nottingham	730	E730BVO	Leyland	B51F	04/88
		City of Nottingham	730	E730BVO	Leyland	B49F	??
LX1123	LX112TL11ZR1R	City of Nottingham	731	E731BVO	Leyland	B51F	04/88
		City of Nottingham	731	E731BVO	Leyland	B49F	11/92
LX1124	LX112TL11ZR1R	City of Nottingham	732	E732BVO	Leyland	B51F	04/88
LX1125	LX112TL11ZR1R	City of Nottingham	733	E733BVO	Leyland	B51F	04/88
		City of Nottingham	733	E733BVO	Leyland	B49F	??
LX1126	LX112TL11ZR1R	City of Nottingham	734	E734BVO	Leyland	B51F	04/88
		City of Nottingham	734	E734BVO	Leyland	B49F	??
LX1127	LX112TL11ZR1R	Safeguard, Guildford		E297OMG	Leyland	B49F	05/88
LX1128	LX112TL11ZR1R	Safeguard, Guildford		E298OMG	Leyland	B49F	05/88
LX1129	LX112TL11ZR1R	Pan Atlas	AB53	E967PME	Leyland	B49F	07/88
		Atlas Bus	LL67	E967PME	Leyland	B49F	07/94
		Crosville Wales	SLC967	E967PME	Leyland	B49F	12/94
LX1130	LX112TL11ZR1R	Pan Atlas	AB56	E968PME	Leyland	B49F	07/88
		Atlas Bus	LL68	E968PME	Leyland	B49F	07/94
		Crosville Wales	SLC968	E968PME	Leyland	B49F	11/94
LX1131	LX112LXCTZR1S	Brighton Transport	47	E447FWV	Leyland	B49F	05/88
LX1132	LX112LXCTZR1S	Brighton Transport	48	E448FWV	Leyland	B49F	05/88
LX1133	LX112LXCTZR1S	Brighton Transport	49	E449FWV	Leyland	B49F	05/88
LX1134	LX112LXCTZR1S	AA Motors (Dodds)		F262WSD	Leyland	B51F	02/90
LX1135	LX112TL11ZR1S	Northern General	4728	F728LRG	Leyland	B51F	02/90
		Sunderland District	4728	F728LRG	Leyland	B51F	05/91
LX1136	B22.L10.Z.R	Leyland Bus		SBS3572Y	Alexander	B53D	00/88
		Singapore BS	3572	SBS3572Y	Alexander	B53D	00/88
LX1137	LX112TL11ZR1S	Northern General	4731	F731LRG	Leyland	B51F	02/89
		Sunderland District	4731	F731LRG	Leyland	B51F	05/91
LX1138	LX112TL11ZR1S	Pan Atlas	AB51	E299OMG	Leyland	B49F	07/89
		Atlas Bus	LL99	E299OMG	Leyland	B49F	07/94
		Crosville Wales	SLC299	E299OMG	Leyland	B49F	12/94
LX1139	LX112TL11ZR1S	Pan Atlas	AB55	E328OMG	Leyland	B49F	07/89
		Atlas Bus	LL28	E328OMG	Leyland	B49F	08/94
		Crosville Wales	SLC328	E328OMG	Leyland	B49F	11/94
LX1140	LX112L10ZR1R	City of Nottingham	735	E735BVO	Leyland	B51F	05/88
		City of Nottingham	735	E735BVO	Leyland	B49F	??
LX1141	LX112L10ZR1R	City of Nottingham	736	E736BVO	Leyland	B51F	05/88
		City of Nottingham	736	E736BVO	Leyland	B49F	??
LX1142	LX112L10ZR1R	City of Nottingham	737	E737BVO	Leyland	B51F	05/88
		City of Nottingham	737	E737BVO	Leyland	B49F	11/92
LX1143	LX112TL11ZR1S	Pan Atlas	AB54	E969PME	Leyland	B49F	07/88
		Atlas Bus	LL69	E969PME	Leyland	B49F	07/94
		Luton & District	417	E969PME	Leyland	B49F	11/94
		The Shires	417	E969PME	Leyland	B49F	04/95
LX1144	LX112TL11ZR1S	Pan Atlas	AB58	E970PME	Leyland	B49F	07/88
		Atlas Bus	LL70	E970PME	Leyland	B49F	08/94
		Luton & District	418	E970PME	Leyland	B49F	11/94
		The Shires	418	E970PME	Leyland	B49F	04/95
LX1145	LX112L10ZR1R	Halton	45	E642VFY	Leyland	B51F	05/88
LX1146	LX112L10ZR1S	Yorkshire Buses	265	E265WUB	Leyland	B49F	05/88
LX1147	LX112L10ZR1S	Yorkshire Buses	266	E266WUB	Leyland	B49F	05/88
LX1148	LX112L10ZR1R	Hedingham	L150	F150LTW	Leyland	B51F	06/88
LX1149	LX112L10ZR1R	Metrobus, Orpington		F80SMC	Leyland	B51F	09/88
LX1150	LX112L10ZR1R	City of Nottingham	738	E738BVO	Leyland	B51F	06/88
		City of Nottingham	738	E738BVO	Leyland	B49F	11/92

LX1151	LX112L10ZR1R	City of Nottingham	739	E739BVO	Leyland	B51F	06/88
		City of Nottingham	739	E739BVO	Leyland	B49F	11/92
LX1152	LX112LXCTZR1S	Simmons, Gt Gonerby		F883SMU	Leyland	B51F	01/89
LX1153	LX112L10ZR1R	Stock, Arlington			Leyland	B51F	08/88
		Volvo, Göteborg			Leyland	B51F	05/89
		Development vehicle, broken up in Sweden					*12/91*
LX1154	LX112L10ZR1R	G & G, Leamington		F661PWK	Leyland	B51F	09/88
		Stagecoach Midland Red	1821	F661PWK	Leyland	B51F	09/94
LX1155	LX112L10ZR1R	G & G, Leamington		F660PWK	Leyland	B51F	09/88
		Stagecoach Midland Red	1820	F660PWK	Leyland	B51F	09/94
LX1156	LX112L10ZR1S	West Riding	267	E267WUB	Leyland	B49F	07/88
LX1157	LX112L10ZR1S	Yorkshire Buses	268	E268WUB	Leyland	B49F	07/88
LX1158	LX112L10ZR1R	Eastern National	1400	E400HWC	Leyland	B49F	07/88
		Thamesway	1400	E400HWC	Leyland	B49F	08/90
LX1159	LX112L10ZR1R	Eastern National	1401	E401HWC	Leyland	B49F	07/88
LX1160	LX112L10ZR1R	Eastern National	1402	F402LTW	Leyland	B49F	07/88
LX1161	LX112L10ZR1R	Merthyr Tydfil	116	E116UTX	Leyland	B51F	07/88
		Yorkshire	351	E116UTX	Leyland	B51F	10/89
LX1162	LX112L10ZR1R	Eastern National	1403	F403LTW	Leyland	B49F	07/88
LX1163	LX112L10ZR1R	Eastern National	1404	F404LTW	Leyland	B49F	07/88
		Thamesway	1404	E404LTW	Leyland	B49F	08/90
LX1164	LX112L10ZR1R	Eastern National	1405	F405LTW	Leyland	B49F	07/88
		Thamesway	1405	E405LTW	Leyland	B49F	08/90
LX1165	LX112L10ZR1S	West Riding	269	E269WUB	Leyland	B49F	07/88
LX1166	LX112L10ZR1S	Yorkshire Buses	270	E270WUB	Leyland	B49F	07/88
LX1167	LX112L10ZR1S	Yorkshire Buses	271	E271WUB	Leyland	B49F	07/88
LX1168	LX112L10ZR1S	Yorkshire Buses	272	F272AWW	Leyland	B49F	08/88
LX1169	LX112L10ZR1R	Topp-Line, Liverpool		F361YTJ	Leyland	B51F	10/88
		PMT	SLC845	F361YTJ	Leyland	B51F	12/94
LX1170	LX112L10ZR1R	Topp-Line, Liverpool		F362YTJ	Leyland	B51F	10/88
		PMT	SLC846	F362YTJ	Leyland	B51F	12/94
LX1171	LX112L10ZR1R	Topp-Line, Liverpool		F363YTJ	Leyland	B51F	10/88
		PMT	SLC847	F363YTJ	Leyland	B51F	12/94
LX1172	LX112L10ZR1R	Eastern National	1406	F406LTW	Leyland	B49F	08/88
		Thamesway	1406	F406LTW	Leyland	B49F	08/90
LX1173	LX112L10ZR1R	Eastern National	1407	F407LTW	Leyland	B49F	08/88
LX1174	LX112L10ZR1R	Eastern National	1408	F408LTW	Leyland	B49F	08/88
LX1175	LX112L10ZR1R	Eastern National	1409	F409LTW	Leyland	B49F	08/88
		Thamesway	1409	F409LTW	Leyland	B49F	08/90
LX1176	LX112L10ZR1R	Eastern National	1410	F410MNO	Leyland	B49F	08/88
		Thamesway	1410	F410MNO	Leyland	B49F	08/90
LX1177	LX112L10ZR1R	Eastern National	1411	F411MNO	Leyland	B49F	08/88
		Thamesway	1411	F411MNO	Leyland	B49F	08/90
LX1178	LX112L10ZR1R	Eastern National	1412	F412MNO	Leyland	B49F	08/88
		Thamesway	1412	F412MNO	Leyland	B49F	08/90
LX1179	LX112L10ZR1R	Eastern National	1413	F413MNO	Leyland	B49F	08/88
LX1180	LX112L10ZR1R	Eastern National	1414	F414MNO	Leyland	B49F	08/88
LX1181	LX112L10ZR1R	Eastern National	.1415	F415MNO	Leyland	B49F	08/88
LX1182	LX112L10ZR1R	Shearings	41	F41ENF	Leyland	B49F	08/88
		Yorkshire Traction	285	F41ENF	Leyland	B49F	08/91
LX1183	LX112L10ZR1R	Shearings	42	F42ENF	Leyland	B49F	08/88
		Yorkshire Traction	286	F42ENF	Leyland	B49F	08/91

Opposite: **Interesting comparison of styles of two bodies from Alexanders are seen opposite. The upper picture shows chassis LX1007 with an Alexander N-type body built for Ulsterbus. It currently operates for Stevensons as 261, HXI3011 in whose livery it is seen here. The Alexander vehicle which operated for Singapore Bus Service featured a body that incorporated much of the PS type style but was fitted with the Leyland front. It is seen in service in Singapore.** *Ken Crawley/Volvo*

F608WBV was registered by Leyland in 1988 and spent several months on demonstration work before being sold to Westbus in 1990. Some three years later, the vehicle has joined the PMT fleet where it carries Crosville names for the Wirral operation. It is seen leaving Heswall bus station in September 1994. *R L Wilson*

LX1184	LX112L10ZR1R	Shearings	43	F43ENF	Leyland	B49F	08/88
		Yorkshire Traction	287	F43ENF	Leyland	B49F	08/91
LX1185	LX112L10ZR1R	Shearings	44	F44ENF	Leyland	B49F	08/88
		Yorkshire Traction	288	F44ENF	Leyland	B49F	08/91
LX1186	LX112L10ZR1R	Eastern National	1416	F416MNO	Leyland	B49F	09/88
LX1187	LX112L10ZR1R	Eastern National	1417	F417MNO	Leyland	B49F	09/88
LX1188	LX112L10ZR1S	Sheffield & District	273	F273AWW	Leyland	B49F	09/88
		Yorkshire	273	F273AWW	Leyland	B49F	10/89
LX1189	LX112L10ZR1S	Sheffield & District	274	F274AWW	Leyland	B49F	09/88
		Yorkshire	274	F274AWW	Leyland	B49F	10/89
LX1190	LX112L10ZR1R	Eastern National	1418	F418MNO	Leyland	B49F	09/88
		Thamesway	1418	F418MNO	Leyland	B49F	08/90
LX1191	LX112L10ZR1R	Eastern National	1419	F419MNO	Leyland	B49F	09/88
		Thamesway	1419	F419MNO	Leyland	B49F	08/90
LX1192	LX112L10ZR1S	West Riding	275	F275AWW	Leyland	B49F	09/88
LX1193	LX112L10ZR1S	West Riding	276	F276AWW	Leyland	B49F	09/88
LX1194	LX112L10ZR1S	Yorkshire	277	F277AWW	Leyland	B49F	09/88
LX1195	LX112L10ZR1R	Volvo Leyland Development			Leyland	B51F	10/88
		Halton	37	J250KWM	Leyland	B51F	10/91
LX1196	LX112L10ZR1R	Topp-Line, Liverpool		F364YTJ	Leyland	B51F	10/88
		PMT	SLC848	F364YTJ	Leyland	B51F	12/94
LX1197	LX112L10ZR1R	Brighton	44	F544LUF	Leyland	B51F	02/89
LX1198	LX112L10ZR1R	Eastern National	1420	F420MJN	Leyland	B49F	09/88
		Thamesway	1420	F420MJN	Leyland	B49F	08/90
LX1199	LX112L10ZR1R	Eastern National	1421	F421MJN	Leyland	B49F	09/88
		Thamesway	1421	F421MJN	Leyland	B49F	08/90

LX1200	LX112L10ZR1R	Eastern National	1422	F422MJN	Leyland	B49F	09/88
		Thamesway	1422	F422MJN	Leyland	B49F	08/90
LX1201	LX112L10ZR1R	Eastern National	1423	F423MJN	Leyland	B49F	09/88
		Thamesway	1423	F423MJN	Leyland	B49F	08/90
LX1202	LX112L10ZR1R	Eastern National	1424	F424MJN	Leyland	B49F	09/88
		Thamesway	1424	F424MJN	Leyland	B49F	08/90
LX1203	LX112L10ZR1R	Eastern National	1425	F425MJN	Leyland	B49F	09/88
LX1204	LX112L10ZR1R	Eastern National	1426	F426MJN	Leyland	B49F	09/88
LX1205	LX112L10ZR1R	Eastern National	1427	F427MJN	Leyland	B49F	09/88
		Eastern National	1427	F427MJN	Leyland	B47F	07/94
LX1206	LX112L10ZR1R	Eastern National	1428	F428MJN	Leyland	B49F	09/88
		Eastern National	1428	F428MJN	Leyland	B47F	07/94
LX1207	LX112L10ZR1R	Eastern National	1429	F429MJN	Leyland	B49F	09/88
		Eastern National	1429	F429MJN	Leyland	B47F	07/94
LX1208	LX112L10ZR1S	West Riding	278	F278AWW	Leyland	B49F	10/88
LX1209	LX112L10ZR1S	Yorkshire	279	F279AWW	Leyland	B49F	10/88
LX1210	LX112L10ZR1S	Yorkshire	280	F280AWW	Leyland	B49F	10/88
LX1211	LX2L56V18	Leyland Bus			Camo	B--F	10/88
LX1212	LX112L10ZR1S	Yorkshire	281	F281AWW	Leyland	B49F	10/88
LX1213	LX112L10ZR1S	West Riding	282	F282AWW	Leyland	B49F	10/88
LX1214	LX112L10ZR1S	West Riding	283	F283AWW	Leyland	B49F	10/88
LX1215	LX112L10ZR1S	Leyland Bus		F608WBV	Leyland	B52F	11/88
		Westbus, Ashford		F608WBV	Leyland	B52F	03/90
		PMT	SLC849	F608WBV	Leyland	B52F	11/93
LX1216	LX112L10ZR1S	Miller, Foxton		F165SMT	Leyland	B51F	01/89
		Metrobus, Orpington		F165SMT	Leyland	B51F	03/91
LX1217	LX112L10ZR1S	Miller, Foxton		F166SMT	Leyland	B51F	01/89
		Metrobus, Orpington		F166SMT	Leyland	B51F	03/91
LX1218	LX112L10ZR1S	The Bee Line	401	F556NJM	Leyland	B52F	10/88
		Wycombe Bus	1401	F556NJM	Leyland	B52F	11/90
LX1219	LX112L10ZR1S	The Bee Line	402	F557NJM	Leyland	B52F	10/88
		Wycombe Bus	1402	F557NJM	Leyland	B52F	11/90
LX1220	LX112L10ZR1S	The Bee Line	403	F558NJM	Leyland	B52F	10/88
		Wycombe Bus	1403	F558NJM	Leyland	B52F	11/90
LX1221	LX112L10ZR1R	Merthyr Tydfil	117	F117XTX	Leyland	B49F	10/88
		Yorkshire	316	F117XTX	Leyland	B51F	10/89
LX1222	LX112L10ZR1S	Miller, Foxton		F167SMT	Leyland	B51F	01/89
		Cambus	310	F167SMT	Leyland	B51F	02/92
LX1223	LX112L10ZR1S	Miller, Foxton		F168SMT	Leyland	B51F	01/89
		Cambus	310	F168SMT	Leyland	B51F	02/92
LX1224	LX112L10ZR1S	Miller, Foxton		F171SMT	Leyland	B51F	05/89
		Cambus	310	F171SMT	Leyland	B51F	02/92
LX1225	LX112L10ZR1S	West Riding	314	F314AWW	Leyland	B49F	02/89
LX1226	LX112L10ZR1S	Busways	101	F101HVK	Leyland	B49F	11/88
LX1227	LX112L10ZR1S	Busways	102	F102HVK	Leyland	B49F	11/88
LX1228	LX112L10ZR1S	Busways	103	F103HVK	Leyland	B49F	11/88
LX1229	LX112L10ZR1S	Busways	104	F104HVK	Leyland	B49F	11/88
LX1230	LX112L10ZR1S	Busways	105	F105HVK	Leyland	B49F	11/88
LX1231	LX112L10ZR1S	Busways	106	F106HVK	Leyland	B49F	11/88
LX1232	LX112L10ZR1S	The Bee Line	404	F559NJM	Leyland	B52F	10/88
		Wycombe Bus	1404	F559NJM	Leyland	B52F	11/90
LX1233	LX112L10ZR1S	The Bee Line	405	F560NJM	Leyland	B52F	10/88
		Wycombe Bus	1405	F560NJM	Leyland	B52F	11/90
LX1234	LX112L10ZR1R	Brighton	38	F538LUF	Leyland	B49F	02/89
LX1235	LX112L10ZR1S	Yorkshire	284	F284AWW	Leyland	B49F	11/88
LX1236	LX112L10ZR1S	Yorkshire	285	F285AWW	Leyland	B49F	11/88
		Sheffield & District	285	F285AWW	Leyland	B49F	05/89
		Yorkshire	285	F285AWW	Leyland	B49F	10/89

LX1237	LX112L10ZR1S	City Bus, Southampton	101	F101FTR	Leyland	B49F	01/89
		Leon, Finningley		F101FTR	Leyland	B49F	12/91
		Leon, Finningley		F101FTR	Leyland	B47F	01/92
LX1238	LX112L10ZR1S	City Bus, Southampton	102	F102FTR	Leyland	B49F	01/89
LX1239	LX112L10ZR1S	City Bus, Southampton	103	F103FTR	Leyland	B49F	01/89
		Leon, Finningley		F103FTR	Leyland	B49F	12/91
		Leon, Finningley		F103FTR	Leyland	B47F	01/92
LX1240	LX112L10ZR1S	West Riding	286	F286AWW	Leyland	B49F	11/88
LX1241	LX112L10ZR1S	Yorkshire	287	F287AWW	Leyland	B49F	11/88
LX1242	LX112L10ZR1S	Sheffield & District	288	F288AWW	Leyland	B49F	11/88
		Yorkshire	288	F288AWW	Leyland	B49F	10/89
LX1243	LX1126LXCTZR1R	Northern General	4722	F722LRG	Leyland	B51F	01/89
		Sunderland District	4722	F722LRG	Leyland	B51F	05/91
LX1244	LX1126LXCTZR1R	Northern General	4727	F727LRG	Leyland	B51F	01/89
		Sunderland District	4727	F727LRG	Leyland	B51F	05/91
LX1245	LX1126LXCTZR1R	Northern General	4723	F723LRG	Leyland	B51F	01/89
		Sunderland District	4723	F723LRG	Leyland	B51F	05/91
LX1246	LX1126LXCTZR1R	Northern General	4724	F724LRG	Leyland	B51F	02/89
		Sunderland District	4724	F724LRG	Leyland	B51F	05/91
LX1247	LX1126LXCTZR1R	Northern General	4725	F725LRG	Leyland	B51F	02/89
		Sunderland District	4725	F725LRG	Leyland	B51F	05/91
LX1248	LX1126LXCTZR1R	Northern General	4726	F726LRG	Leyland	B51F	02/89
		Sunderland District	4726	F726LRG	Leyland	B51F	05/91
LX1249	LX1126LXCTZR1R	Northern General	4732	F732LRG	Leyland	B51F	02/89
		Sunderland District	4732	F732LRG	Leyland	B51F	05/91
LX1250	LX1126LXCTZR1R	Northern General	4730	F730LRG	Leyland	B51F	02/89
		Sunderland District	4730	F730LRG	Leyland	B51F	05/91
LX1251	LX1126LXCTZR1R	Northern General	4729	F729LRG	Leyland	B51F	02/89
		Sunderland District	4729	F729LRG	Leyland	B51F	05/91
LX1252	LX112L10ZR1S	West Riding	289	F289AWW	Leyland	B49F	12/88
LX1253	LX112L10ZR1S	Sheffield & District	290	F290AWW	Leyland	B49F	12/88
		West Riding	290	F290AWW	Leyland	B49F	10/89
LX1254	LX112L10ZR1S	Yorkshire	291	F291AWW	Leyland	B49F	12/88
LX1255	LX112L10ZR1S	West Riding	292	F292AWW	Leyland	B49F	12/88
LX1256	LX112L10ZR1S	Sheffield & District	293	F293AWW	Leyland	B49F	12/88
		West Riding	293	F293AWW	Leyland	B49F	10/89
LX1257	LX112L10ZR1S	West Riding	294	F294AWW	Leyland	B49F	12/88
LX1258	LX112L10ZR1S	Yorkshire	295	F295AWW	Leyland	B49F	12/88
LX1259	LX112L10ZR1S	Yorkshire	296	F296AWW	Leyland	B49F	12/88
LX1260	LX112L10ZR1R	Brighton	46	F546LUF	Leyland	B51F	02/89
LX1261	LX112L10ZR1R	Brighton	45	F545LUF	Leyland	B51F	02/89
LX1262	LX112L10ZR1R	The Wright Company		F258GWJ	Leyland	B51F	04/89
		Stevensons	102	F258GWJ	Leyland	B51F	11/93
LX1263	LX112L10ZR1S	West Yorkshire Buses	1203	F203MBT	Leyland	B51F	04/89
		Harrogate & District	1203	F203MBT	Leyland	B51F	08/89
		York City & District	3	F203MBT	Leyland	B51F	04/90
		Harrogate & District	373	F203MBT	Leyland	B51F	07/90
		Keighley & District	209	F203MBT	Leyland	B51F	04/92
		Sovereign Bus & Coach	213	F203MBT	Leyland	B51F	08/93
LX1264	LX112L10ZR1S	West Yorkshire Buses	1204	F204MBT	Leyland	B51F	04/89
		Harrogate & District	1204	F204MBT	Leyland	B51F	08/89
		York City & District	4	F204MBT	Leyland	B51F	04/90
		Harrogate & District	374	F204MBT	Leyland	B51F	07/90
		Keighley & District	210	F204MBT	Leyland	B51F	04/92
		Sovereign Bus & Coach	214	F204MBT	Leyland	B51F	08/93

Opposite: **Representing the independents who operate the Lynx are another view of F608WBV, this time in Westbus livery, and F101RTR seen here in the livery of Leon of Finningley. The latter example was new to City Bus of Southampton who sold two examples to Leon during 1991.**
Keith Grimes/Mike Fowler

LX1265	LX112L10ZR1S	West Yorkshire Buses	1205	F205MBT	Leyland	B51F	04/89
		Harrogate & District	1205	F205MBT	Leyland	B51F	08/89
		York City & District	5	F205MBT	Leyland	B51F	04/90
		Harrogate & District	375	F205MBT	Leyland	B51F	07/90
		Keighley & District	211	F205MBT	Leyland	B51F	04/92
		Sovereign Bus & Coach	215	F205MBT	Leyland	B51F	08/93
LX1266	LX112L10ZR1S	Busways	107	F107HVK	Leyland	B49F	01/89
LX1267	LX112L10ZR1R	Halton	8	F687YWM	Leyland	B51F	12/88
LX1268	LX112L10ZR1R	Northern General	4733	F733LRG	Leyland	B51F	02/89
		Sunderland & District	4733	F733LRG	Leyland	B51F	05/91
LX1269	LX112L10ZR1R	Brighton	92	G992VWV	Leyland	B51F	08/90
LX1270	LX112L10ZR1R	Harrogate & District	381	G381MWU	Leyland	DP47F	04/90
LX1271	LX112L10ZR1R	Harrogate & District	382	G382MWU	Leyland	DP47F	04/90
LX1272	LX112L10ZR1R	Harrogate & District	383	G383MWU	Leyland	DP47F	04/90
LX1273	LX112L10ZR1R	Harrogate & District	384	G384MWU	Leyland	DP47F	04/90
		Sovereign Bus & Coach	284	G384MWU	Leyland	DP47F	08/95
LX1274	LX112L10ZR1R	Halton	11	H34HBG	Leyland	B51F	02/91
LX1275	LX112L10ZR1R	Isle of Man	1	BMN401T	Leyland	B51F	01/90
LX1276	LX112L10ZR1R	Isle of Man	2	BMN402T	Leyland	B51F	01/90
LX1277	LX112L10ZR1R	Westbus, Ashford		G936VRY	Leyland	B51F	03/90
		PMT	SLC850	G936VRY	Leyland	B51F	11/93
LX1278	LX112L10ZR1R	Brighton	93	G993VWV	Leyland	B51F	08/90
LX1279	LX112L10ZR1R	Brighton	94	G994VWV	Leyland	B51F	08/90
LX1280	LX112L10ZR1R	Brighton	95	G995VWV	Leyland	B51F	08/90
LX1281	LX112L10ZR1R	Brighton	96	G996VWV	Leyland	B51F	08/90
LX1282	LX112L10ZR1S	West Riding	297	F297AWW	Leyland	B49F	01/89
LX1283	LX112L10ZR1S	West Riding	298	F298AWW	Leyland	B49F	01/89
LX1284	LX112L10ZR1S	West Riding	299	F299AWW	Leyland	B49F	02/89
LX1285	LX112L10ZR1S	West Riding	300	F300AWW	Leyland	B49F	01/89
LX1286	LX112L10ZR1S	Sheffield & District	301	F301AWW	Leyland	B49F	02/89
		West Riding	301	F301AWW	Leyland	B49F	10/89
LX1287	LX112L10ZR1S	West Riding	302	F302AWW	Leyland	B49F	01/89
LX1288	LX112L10ZR1S	Yorkshire Buses	303	F303AWW	Leyland	B49F	02/89
LX1289	LX112L10ZR1S	Yorkshire Buses	304	F304AWW	Leyland	B49F	02/89
LX1290	LX112L10ZR1R	Merthyr Tydfil	118	F118XTX	Leyland	B51F	01/89
		Yorkshire Buses	317	F118XTX	Leyland	B51F	10/89
LX1291	LX112L10ZR1R	Hillingdon LB	LX1	F101GRM	Leyland	B47F	12/88
		London Buses	LX1	F101GRM	Leyland	B47F	09/91
		CentreWest	LX1	F101GRM	Leyland	B47F	09/94
LX1292	LX112L10ZR1R	Hillingdon LB	LX2	F102GRM	Leyland	B47F	12/88
		London Buses	LX2	F102GRM	Leyland	B47F	09/91
		CentreWest	LX2	F102GRM	Leyland	B47F	09/94
LX1293	LX112L10ZR1R	Shearings	45	F45ENF	Leyland	B49F	12/88
		Maidstone & District	3045	F45ENF	Leyland	B49F	12/91
LX1294	LX112L10ZR1R	Shearings	46	F46ENF	Leyland	B49F	12/88
		Maidstone & District	3046	F46ENF	Leyland	B49F	12/91
LX1295	LX112L10ZR1R	Shearings	47	F47ENF	Leyland	B49F	12/88
		Maidstone & District	3047	F47ENF	Leyland	B49F	12/91
LX1296	LX112L10ZR1R	Shearings	48	F48ENF	Leyland	B49F	12/88
		Maidstone & District	3048	F48ENF	Leyland	B49F	12/91
LX1297	LX112L10ZR1R	City of Nottingham	740	F740HRC	Leyland	B51F	01/89
		City of Nottingham	740	F740HRC	Leyland	B49F	11/92
LX1298	LX112L10ZR1R	City of Nottingham	741	F741HRC	Leyland	B51F	01/89
LX1299	LX112L10ZR1R	City of Nottingham	742	F742HRC	Leyland	B51F	01/89
LX1300	LX112L10ZR1R	City of Nottingham	743	F743HRC	Leyland	B51F	01/89
LX1301	LX112L10ZR1R	City of Nottingham	744	F744HRC	Leyland	B51F	01/89

LX1302	LX112L10ZR1S	West Yorkshire Buses	1206	F206MBT	Leyland	B52F	04/89
		Harrogate & District	1206	F206MBT	Leyland	B52F	08/89
		York City & District	6	F206MBT	Leyland	B52F	04/90
		Harrogate & District	376	F206MBT	Leyland	B52F	07/90
		Keighley & District	212	F206MBT	Leyland	B52F	04/92
		Sovereign Bus & Coach	216	F206MBT	Leyland	B52F	09/93
LX1303	LX112L10ZR1S	West Yorkshire Buses	1207	F207MBT	Leyland	B52F	04/89
		Harrogate & District	1207	F207MBT	Leyland	B52F	08/89
		York City & District	7	F207MBT	Leyland	B52F	04/90
		Harrogate & District	377	F207MBT	Leyland	B52F	07/90
		Sovereign Bus & Coach	217	F207MBT	Leyland	B52F	01/93
LX1304	LX112L10ZR1S	West Yorkshire Buses	1208	F208MBT	Leyland	B52F	04/89
		Harrogate & District	1208	F208MBT	Leyland	B52F	08/89
		York City & District	8	F208MBT	Leyland	B52F	04/90
		Harrogate & District	378	F208MBT	Leyland	B52F	07/90
		Sovereign Bus & Coach	219	F208MBT	Leyland	B52F	01/93
LX1305	LX112L10ZR1R	The Wright Company		F259GWJ	Leyland	B51F	04/89
		Redby Travel, Sunderland		F259GWJ	Leyland	B51F	12/93
LX1306	LX112L10ZR1R	Rhodeservice, Yeadon	314	G261LUG	Leyland	B51F	11/89
		Quickstep		G261LUG	Leyland	B51F	04/94
		Brewers	507	G261LUG	Leyland	B51F	04/94
LX1307	LX112L10ZR1R	Whitelaw, Stonehouse		G472PGE	Leyland	B51F	09/89
		Leon, Finningley	135	G472PGE	Leyland	B51F	08/93
		UniversityBus, Hatfield		G472PGE	Leyland	B51F	10/93
LX1308	LX112L10ZR1R	Citybus, Southampton	112	G112XOW	Leyland	DP47F	03/90
LX1309	LX112L10ZR1R	Citybus, Southampton	113	G113XOW	Leyland	DP47F	03/90
LX1310	LX112L10ZR1R	Cumberland	251	F251JRM	Leyland	B51F	06/89
LX1311	LX112L10ZR1R	Cumberland	252	F252JRM	Leyland	B51F	06/89
LX1312	LX112L10ZR1R	Cumberland	253	F253KAO	Leyland	B51F	06/89
LX1313	LX112L10ZR1R	Luton & District	400	F400PUR	Leyland	B51F	06/89
		The Shires	400	F400PUR	Leyland	B49F	04/95
LX1314	LX112L10ZR1R	Keighley & District	201	G293KWY	Leyland	B49F	11/89
LX1315	LX112L10ZR1R	Keighley & District	202	G294KWY	Leyland	B49F	11/89
LX1316	LX112L10ZR1R	Whitelaw, Stonehouse		G473PGE	Leyland	B51F	09/89
LX1317	LX112L10ZR1R	Volvo Bus		G49CVC	Leyland	B51F	02/90
		Loan to West Riding		G49CVC	Leyland	B51F	02/90
		Yorkshire	351	G49CVC	Leyland	B51F	01/91
LX1318	LX112L10ZR1R	The Wright Company		H408YMA	Leyland	B51F	12/90
		Stevensons	106	H408YMA	Leyland	B51F	01/94
LX1319	LX112L10ZR1R	Halton	9	F81STB	Leyland	B51F	01/89
LX1320	LX112L10ZR1R	Keighley & District	203	G295KWY	Leyland	B49F	12/89
LX1321	LX112L10ZR1R	Keighley & District	204	G296KWY	Leyland	B49F	12/89
LX1322	LX112L10ZR1R	Keighley & District	205	G297KWY	Leyland	B49F	12/89
		Harrogate & District	385	G297KWY	Leyland	B49F	09/94
		Keighley & District	205	G297KWY	Leyland	B49F	05/95
LX1323	LX112L10ZR1R	Keighley & District	206	G298KWY	Leyland	B49F	12/89
LX1324	LX112L10ZR1R	Keighley & District	207	G299KWY	Leyland	B49F	12/89
LX1325	LX112L10ZR1R	Keighley & District	208	G300KWY	Leyland	B49F	12/89
LX1326	LX112L10ZR1R	Halton	54	G473DHF	Leyland	B51F	01/90
LX1327	LX112L10ZR1R	Halton	27	G474DHF	Leyland	B51F	01/90
LX1328	LX112L10ZR1R	Luton & District	401	F401PUR	Leyland	B51F	06/89
		The Shires	401	F401PUR	Leyland	B51F	04/95
LX1329	LX112L10ZR1R	Luton & District	402	F402PUR	Leyland	B51F	06/89
		The Shires	402	F402PUR	Leyland	B51F	04/95
LX1330	LX112L10ZR1R	Luton & District	403	F403PUR	Leyland	B51F	06/89
		The Shires	403	F403PUR	Leyland	B51F	04/95
LX1331	LX112L10ZR1R	AA Motors(Dodds)		F85XCS	Leyland	B51F	02/89
LX1332	LX112L10ZR1R	Luton & District	404	F404PUR	Leyland	B51F	06/89
		The Shires	404	F404PUR	Leyland	B51F	04/95

LX1333	LX112L10ZR1R	Halton	14	F520AEM	Leyland	B51F	02/89
LX1334	LX112L10ZR1R	Cleveland Transit	1	F601UVN	Leyland	B49F	02/89
LX1335	LX112L10ZR1R	Cleveland Transit	2	F602UVN	Leyland	B49F	02/89
LX1336	LX112L10ZR1S	Busways	115	F115HVK	Leyland	B49F	02/89
LX1337	LX112L10ZR1S	Busways	116	F116HVK	Leyland	B49F	02/89
LX1338	LX112L10ZR1S	Busways	117	F117HVK	Leyland	B49F	02/89
LX1339	LX112L10ZR1S	Busways	118	F118HVK	Leyland	B49F	02/89
LX1340	LX112L10ZR1S	Busways	119	F119HVK	Leyland	B49F	02/89
LX1341	LX112L10ZR1S	Busways	108	F108HVK	Leyland	B49F	02/89
LX1342	LX112L10ZR1S	Busways	109	F109HVK	Leyland	B49F	02/89
LX1343	LX112L10ZR1S	Busways	110	F110HVK	Leyland	B49F	02/89
LX1344	LX112L10ZR1S	Busways	111	F111HVK	Leyland	B49F	02/89
LX1345	LX112L10ZR1S	Busways	112	F112HVK	Leyland	B49F	02/89
LX1346	LX112L10ZR1S	Busways	113	F113HVK	Leyland	B49F	02/89
LX1347	LX112L10ZR1S	Busways	114	F114HVK	Leyland	B49F	02/89
LX1348	LX112L10ZR1R	Cleveland Transit	3	F603UVN	Leyland	B49F	02/89
LX1349	LX112L10ZR1R	Cleveland Transit	4	F604UVN	Leyland	B49F	02/89
LX1350	LX112L10ZR1R	Stevensons	61	F61PRE	Leyland	B49F	02/89
LX1351	LX112L10ZR1R	Cardiff Bus	231	F231CNY	Leyland	B51F	03/89
LX1352	LX112L10ZR1R	Cardiff Bus	232	F232CNY	Leyland	B51F	03/89
LX1353	LX112L10ZR1R	Cardiff Bus	233	F233CNY	Leyland	B51F	03/89
LX1354	LX112L10ZR1R	Cardiff Bus	234	F234CNY	Leyland	B51F	03/89
LX1355	LX112L10ZR1R	Cardiff Bus	235	F235CNY	Leyland	B51F	03/89
LX1356	LX112L10ZR1R	Cardiff Bus	236	F236CNY	Leyland	B51F	03/89
LX1357	LX112L10ZR1R	Cleveland Transit	5	F605UVN	Leyland	B49F	02/89
LX1358	LX112L10ZR1R	Cleveland Transit	6	F606UVN	Leyland	B49F	02/89
LX1359	LX112L10ZR1R	Cleveland Transit	7	F607UVN	Leyland	B49F	03/89
LX1360	LX112L10ZR1R	Cleveland Transit	8	F608UVN	Leyland	B49F	03/89
LX1361	LX112L10ZR1R	Cleveland Transit	9	F609UVN	Leyland	B49F	03/89
LX1362	LX112L10ZR1R	Cleveland Transit	10	F610UVN	Leyland	B49F	03/89
LX1363	LX112L10ZR1R	Preston Bus	10	F210YHG	Leyland	B47F	03/89
LX1364	LX112L10ZR1R	Preston Bus	11	F211YHG	Leyland	B47F	03/89
LX1365	LX112L10ZR1R	Chesterfield	66	F66FKW	Leyland	B49F	04/89
		Crosville Wales	SLC66	F66FKW	Leyland	B49F	03/95
LX1366	LX112L10ZR1R	Chesterfield	67	F67FKW	Leyland	B49F	04/89
		Crosville Wales	SLC67	F67FKW	Leyland	B49F	03/95
LX1367	LX112L10ZR1R	Chesterfield	68	F68FKW	Leyland	B49F	04/89
		Crosville Wales	SLC68	F68FKW	Leyland	B49F	03/95
LX1368	LX112L10ZR1R	Chesterfield	69	F69FKW	Leyland	B49F	04/89
		Crosville Wales	SLC69	F69FKW	Leyland	B49F	04/95
LX1369	LX112L10ZR1R	Chesterfield	70	F70FKW	Leyland	B49F	04/89
		Crosville Wales	SLC70	F70FKW	Leyland	B49F	04/95
LX1370	LX112L10ZR1S	Busways	120	F120HVK	Leyland	B49F	03/89
LX1371	LX112L10ZR1S	Busways	121	F121HVK	Leyland	B49F	03/89
LX1372	LX112L10ZR1S	Busways	122	F122HVK	Leyland	B49F	03/89
LX1373	LX112L10ZR1S	Busways	123	F123HVK	Leyland	B49F	03/89
LX1374	LX112L10ZR1S	Busways	124	F124HVK	Leyland	B49F	03/89
LX1375	LX112L10ZR1S	Busways	125	F125HVK	Leyland	B49F	03/89
LX1376	LX112L10ZR1R	Preston Bus	12	F212YHG	Leyland	DP43F	03/89
LX1377	LX112L10ZR1R	Preston Bus	13	F213YHG	Leyland	DP43F	03/89

Opposite: **Two significant orders for the Lynx came from metropolitan operators. Busways on Tyneside took the type as its main single deck during 1989 while West Midlands Travel ordered 250 for delivery in 1989 following from the initial half-dozen 1986 examples. Interestingly, West Midlands Travel have a letter on the door sides to indicate to the visually handicapped the vehicle type. For the Lynx this is 'L' and was attached to the LX1686 and LX1687 which were diverted to Australia. For a time these operated in the livery shown here, but with Action fleet names.**
Philip Stephenson/Keith Grimes

LX1308 for CityBus, Southampton, was one of a pair fitted with high-back seating. Winchester bus station is the setting for this view of 112, G112XOW taken shortly after delivery in 1990. *Mark Jameson*

LX1378	LX2R11C15Z4R	CityBus Southampton	104	G104WRV	Leyland	B49F	02/90
LX1379	LX2R11C15Z4R	CityBus Southampton	105	G105WRV	Leyland	B49F	02/90
LX1380	LX2R11C15Z4R	Cardiff	237	F237CNY	Leyland	DP47F	03/89
LX1381	LX2R11C15Z4R	Cardiff	238	F238CNY	Leyland	DP47F	03/89
LX1382	LX2R11C15Z4R	Cardiff	239	F239CNY	Leyland	DP47F	03/89
LX1383	LX2R11C15Z4R	Cardiff	240	F240CNY	Leyland	DP47F	03/89
LX1384	LX112L10ZR1R	Halton	15	F521AEM	Leyland	B51F	04/89
LX1385	LX2R11C15Z4R	CityBus, Southampton	106	G106WRV	Leyland	B49F	02/90
LX1386	LX2R11C15Z4R	CityBus, Southampton	107	G107WRV	Leyland	B49F	02/90
		Dismantled follwoing Accident					03/93
LX1387	LX2R11C15Z4R	CityBus, Southampton	108	G108WRV	Leyland	B49F	02/90
LX1388	LX2R11C15Z4R	CityBus, Southampton	109	G109WRV	Leyland	B49F	02/90
LX1389	LX2R11C15Z4R	CityBus, Southampton	110	G110WRV	Leyland	B49F	02/90
LX1390	LX2R11C15Z4R	City Line	1600	F600RTC	Leyland	B49F	02/90
LX1391	LX2R11C15Z4R	CityBus, Southampton	111	G111XOW	Leyland	B49F	02/90
LX1392	LX2R11C15Z4R	Volvo Bus		G149CHP	Leyland	B51F	02/90
		Loan to West Riding		G149CHP	Leyland	B51F	02/90
		Yorkshire	350	G149CHP	Leyland	B51F	01/91
LX1393	LX2R11C15Z4R	Volvo Bus		G148CHP	Leyland	B51F	02/90
		Loan to West Riding		G148CHP	Leyland	B51F	02/90
		West Riding	349	G148CHP	Leyland	B51F	01/91
LX1394	LX2R11C15Z4R	Preston Bus	23	H23YBV	Leyland	DP45F	11/90
LX1395	LX2R11C15Z4R	Preston Bus	24	H24YBV	Leyland	DP45F	11/90
LX1396	LX2R11C15Z4R	Preston Bus	26	H26YBV	Leyland	DP45F	11/90
LX1397	LX2R11C15Z4R	Preston Bus	27	H27YBV	Leyland	DP45F	11/90
LX1398	LX2R11C15Z4R	Volvo Bus		G542CAC	Leyland	B51F	02/90
		Loan to West Riding		G542CAC	Leyland	B51F	02/90
		West Riding	348	G542CAC	Leyland	B51F	01/91
LX1399	LX2R11C15Z4R	Preston Bus	28	H28YBV	Leyland	DP45F	12/90
LX1400	LX2R11C15Z4R	Halton	55	G803EKA	Leyland	B51F	05/90

Transit operate thirty Lynx which consists of twenty of the original model and ten Lynx IIs. Pictured here is 14, G614CEF built in 1989 as LX1416. *David Cole*

LX1401	LX2R11C15Z4R	Preston Bus	29	H29YBV	Leyland	DP45F	12/90
LX1402	LX2R11C15Z4R	Cardiff Bus	249	G249HUH	Leyland	B51F	01/90
LX1403	LX2R11C15Z4R	Cardiff Bus	250	G250HUH	Leyland	B51F	01/90
LX1404	LX112L10ZR1S	West Riding	305	F305AWW	Leyland	B49F	03/89
LX1405	LX112L10ZR1S	West Riding	306	F306AWW	Leyland	B49F	03/89
LX1406	LX112L10ZR1S	West Riding	307	F307AWW	Leyland	B49F	03/89
LX1407	LX112L10ZR1S	West Riding	308	F308AWW	Leyland	B49F	03/89
LX1408	LX112L10ZR1S	West Riding	309	F309AWW	Leyland	B49F	04/89
LX1409	LX112L10ZR1S	West Riding	310	F310AWW	Leyland	B49F	04/89
LX1410	LX112L10ZR1S	West Riding	311	F311AWW	Leyland	B49F	04/89
LX1411	LX112L10ZR1S	West Riding	312	F312AWW	Leyland	B49F	04/89
LX1412	LX112L10ZR1S	West Riding	313	F313AWW	Leyland	B49F	04/89
LX1413	LX2R11C15Z4R	Cleveland	11	G611GEF	Leyland	B51F	12/89
LX1414	LX2R11C15Z4R	Cleveland	12	G612GEF	Leyland	B51F	12/89
LX1415	LX2R11C15Z4R	Cleveland	13	G613GEF	Leyland	B51F	12/89
LX1416	LX2R11C15Z4R	Cleveland	14	G614GEF	Leyland	B51F	12/89
LX1417	LX2R11C15Z4R	Cleveland	15	G615GEF	Leyland	B51F	12/89
LX1418	LX2R11C15Z4R	Cleveland	16	G616GEF	Leyland	B51F	12/89
LX1419	LX2R11C15Z4R	Cleveland	17	G617GEF	Leyland	B51F	12/89
LX1420	LX2R11C15Z4R	Cleveland	18	G618GEF	Leyland	B51F	12/89
LX1421	LX2R11C15Z4R	Cleveland	19	G619GEF	Leyland	B51F	12/89
LX1422	LX2R11C15Z4R	Cleveland	20	G620GEF	Leyland	B51F	12/89
LX1423	LX2R11C15Z4R	Cardiff Bus	251	G251HUH	Leyland	B49F	01/90
LX1424	LX2R11C15Z4R	Halton	56	H542FWM	Leyland	B51F	06/90
LX1425	LX2R11C15Z4R	Leyland Bus		F74DCW	Leyland	B49F	07/89
		Cynon Valley	54	F74DCW	Leyland	B49F	02/90
		Red & White	398	F74DCW	Leyland	DP47F	08/92

LX1426	LX2R11C15Z4R	Cardiff Bus	241	F241CNY	Leyland	B49F	05/89
LX1427	LX2R11C15Z4R	Cardiff Bus	242	F242CNY	Leyland	B49F	05/89
LX1428	LX2R11C15Z4R	Cardiff Bus	243	F243CNY	Leyland	B49F	05/89
LX1429	LX2R11C15Z4R	Cardiff Bus	244	F244CNY	Leyland	B49F	05/89
LX1430	LX2R11C15Z4R	Cardiff Bus	245	F245CNY	Leyland	B49F	05/89
LX1431	LX2R11C15Z4R	Cardiff Bus	246	F246CNY	Leyland	B49F	05/89
LX1432	LX2R11C15Z4R	Cardiff Bus	247	F247CNY	Leyland	B49F	05/89
LX1433	LX2R11C15Z4R	Cardiff Bus	248	F248CNY	Leyland	B49F	05/89
LX1434	LX2R11C15Z4R	Cardiff Bus	252	G252HUH	Leyland	B49F	01/90
LX1435	LX2R11C15Z4R	Cardiff Bus	253	G253HUH	Leyland	B49F	01/90
LX1436	LX2R11C15Z4R	West Midlands Travel	1067	F67DDA	Leyland	B49F	07/89
LX1437	LX2R11C15Z4R	Cardiff Bus	254	G254HUH	Leyland	B49F	01/90
LX1438	LX2R11C15Z4R	Cardiff Bus	255	G255HUH	Leyland	B49F	01/90
LX1439	LX2R11C15Z4R	Cardiff Bus	256	G256HUH	Leyland	B49F	01/90
LX1440	LX112L10ZR1R	Halton	16	F895BKF	Leyland	B51F	06/89
LX1441	LX2R11C15Z4R	Cardiff Bus	257	G257HUH	Leyland	B49F	01/90
LX1442	LX2R11C15Z4R	Cardiff Bus	258	G258HUH	Leyland	B49F	01/90
LX1443	LX2R11C15Z4R	Cardiff Bus	259	G259HUH	Leyland	B49F	01/90
LX1444	LX2R11C15Z4R	City Line	1601	F601RTC	Leyland	B49F	06/89
LX1445	LX2R11C15Z4R	City Line	1602	F602RTC	Leyland	B49F	06/89
LX1446	LX2R11C15Z4R	City Line	1603	F603RTC	Leyland	B49F	06/89
LX1447	LX2R11C15Z4R	City Line	1604	F604RTC	Leyland	B49F	06/89
LX1448	LX2R11C15Z4R	City Line	1605	F605RTC	Leyland	B49F	06/89
LX1449	LX2R11C15Z4R	City Line	1606	F606RTC	Leyland	B49F	06/89
LX1450	LX2R11C15Z4R	City Line	1607	F607RTC	Leyland	B49F	06/89
LX1451	LX2R11C15Z4R	City Line	1608	F608RTC	Leyland	B49F	06/89
LX1452	LX2R11C15Z4R	City Line	1609	F609RTC	Leyland	B49F	06/89
LX1453	LX2R11C15Z4R	City Line	1610	F610RTC	Leyland	B49F	06/89
LX1454	LX2R11C15Z4R	City Line	1611	F611RTC	Leyland	B49F	06/89
LX1455	LX2R11C15Z4R	City Line	1612	F612RTC	Leyland	B49F	06/89
LX1456	LX2R11C15Z4R	City Line	1613	F613RTC	Leyland	B49F	06/89
LX1457	LX2R11C15Z4R	City Line	1614	F614RTC	Leyland	B49F	06/89
LX1458	LX2R11C15Z4R	City Line	1615	F615RTC	Leyland	B49F	06/89
LX1459	LX2R11C15Z4R	City Line	1616	F616RTC	Leyland	B49F	06/89
LX1460	LX2R11C15Z4R	City Line	1617	F617RTC	Leyland	B49F	06/89
LX1461	LX2R11C15Z4R	City Line	1618	F618RTC	Leyland	B49F	06/89
LX1462	LX2R11C15Z4R	City Line	1619	F619RTC	Leyland	B49F	06/89
LX1463	LX2R11C15Z4R	City Line	1620	F620RTC	Leyland	B49F	06/89
LX1464	LX2R11C15Z4R	City Line	1621	F621RTC	Leyland	B49F	06/89
LX1465	LX2R11C15Z4R	City Line	1622	F622RTC	Leyland	B49F	06/89
LX1466	LX2R11C15Z4R	City Line	1623	F623RTC	Leyland	B49F	06/89
LX1467	LX2R11C15Z4R	West Midlands Travel	1068	F68DDA	Leyland	B49F	06/89
LX1468	LX2R11C15Z4R	West Midlands Travel	1069	F69DDA	Leyland	B49F	06/89
LX1469	LX2R11C15Z4R	West Midlands Travel	1070	F70DDA	Leyland	B49F	06/89
LX1470	LX2R11C15Z4R	West Midlands Travel	1071	F71DDA	Leyland	B49F	06/89
LX1471	LX2R11C15Z4R	West Midlands Travel	1072	F72DDA	Leyland	B49F	06/89
LX1472	LX2R11C15Z4R	West Midlands Travel	1073	F73DDA	Leyland	B49F	06/89
LX1473	LX2R11C15Z4R	West Midlands Travel	1074	F74DDA	Leyland	B49F	06/89
LX1474	LX2R11C15Z4R	West Midlands Travel	1075	F75DDA	Leyland	B49F	06/89
LX1475	LX2R11C15Z4R	West Midlands Travel	1076	F76DDA	Leyland	B49F	06/89
LX1476	LX2R11C15Z4R	West Midlands Travel	1077	F77DDA	Leyland	B49F	06/89
LX1477	LX2R11C15Z4R	West Midlands Travel	1078	F78DDA	Leyland	B49F	06/89
LX1478	LX2R11C15Z4S	London Buses	LX3	G73UYV	Leyland	B49F	08/89
		London United	LX3	G73UYV	Leyland	B49F	11/94
LX1479	LX2R11C15Z4R	City Line	1624	F624RTC	Leyland	B49F	07/89
LX1480	LX2R11C15Z4R	City Line	1625	F625RTC	Leyland	B49F	07/89
LX1481	LX2R11C15Z4R	City Line	1626	F626RTC	Leyland	B49F	07/89

LX1482	LX2R11C15Z4R	West Midlands Travel	1079	G79EOG	Leyland	B49F	08/89
LX1483	LX2R11C15Z4R	West Midlands Travel	1080	G80EOG	Leyland	B49F	08/89
LX1484	LX2R11C15Z4R	West Midlands Travel	1081	G81EOG	Leyland	B49F	08/89
LX1485	LX2R11C15Z4R	West Midlands Travel	1082	G82EOG	Leyland	B49F	08/89
LX1486	LX2R11C15Z4R	West Midlands Travel	1083	G83EOG	Leyland	B49F	08/89
LX1487	LX2R11C15Z4R	West Midlands Travel	1084	G84EOG	Leyland	B49F	08/89
LX1488	LX2R11C15Z4R	West Midlands Travel	1085	G85EOG	Leyland	B49F	08/89
LX1489	LX2R11C15Z4R	West Midlands Travel	1086	G86EOG	Leyland	B49F	08/89
LX1490	LX2R11C15Z4R	City Line	1627	F627RTC	Leyland	B49F	07/89
LX1491	LX2R11C15Z4R	City Line	1628	F628RTC	Leyland	B49F	07/89
LX1492	LX2R11C15Z4R	City Line	1629	F629RTC	Leyland	B49F	07/89
LX1493	LX2R11C15Z4R	City Line	1630	F630RTC	Leyland	B49F	07/89
LX1494	LX2R11C15Z4R	City Line	1631	F631RTC	Leyland	B49F	07/89
LX1495	LX2R11C15Z4R	City Line	1632	F632RTC	Leyland	B49F	07/89
LX1496	LX2R11C15Z4R	West Midlands Travel	1087	G87EOG	Leyland	B49F	08/89
LX1497	LX2R11C15Z4R	West Midlands Travel	1088	G88EOG	Leyland	B49F	08/89
LX1498	LX2R11C15Z4R	West Midlands Travel	1089	G89EOG	Leyland	B49F	08/89
LX1499	LX2R11C15Z4R	West Midlands Travel	1090	G90EOG	Leyland	B49F	08/89
LX1500	LX2R11C15Z4R	West Midlands Travel	1091	G91EOG	Leyland	B49F	08/89
LX1501	LX2R11C15Z4R	West Midlands Travel	1092	G92EOG	Leyland	B49F	08/89
LX1502	LX2R11C15Z4R	West Midlands Travel	1093	G93EOG	Leyland	B49F	08/89
LX1503	LX2R11C15Z4R	West Midlands Travel	1094	G94EOG	Leyland	B49F	08/89
LX1504	LX2R11C15Z4S	Boro'line, Maidstone	801	G34VME	Leyland	B49F	09/89
		Kentish Bus	401	G34VME	Leyland	B49F	02/92
LX1505	LX2R11C15Z4S	Boro'line, Maidstone	802	G35VME	Leyland	B49F	09/89
		Kentish Bus	402	G35VME	Leyland	B49F	02/92
LX1506	LX2R11C15Z4S	Boro'line, Maidstone	803	G36VME	Leyland	B49F	09/89
		Kentish Bus	403	G36VME	Leyland	B49F	02/92
LX1507	LX2R11C15Z4S	Boro'line, Maidstone	804	G37VME	Leyland	B49F	09/89
		Kentish Bus	404	G37VME	Leyland	B49F	02/92
LX1508	LX2R11C15Z4S	Boro'line, Maidstone	805	G38VME	Leyland	B49F	09/89
		Kentish Bus	405	G38VME	Leyland	B49F	02/92
LX1509	LX2R11C15Z4S	Boro'line, Maidstone	806	G39VME	Leyland	B49F	09/89
		Kentish Bus	406	G39VME	Leyland	B49F	02/92
LX1510	LX2R11C15Z4R	Halton	17	G221DKA	Leyland	B51F	09/89
LX1511	LX2R11C15Z4R	West Midlands Travel	1095	G95EOG	Leyland	B49F	09/89
LX1512	LX2R11C15Z4R	West Midlands Travel	1096	G96EOG	Leyland	B49F	09/89
LX1513	LX2R11C15Z4R	West Midlands Travel	1097	G97EOG	Leyland	B49F	09/89
LX1514	LX2R11C15Z4R	West Midlands Travel	1098	G98EOG	Leyland	B49F	09/89
LX1515	LX2R11C15Z4R	West Midlands Travel	1099	G99EOG	Leyland	B49F	09/89
LX1516	LX2R11C15Z4R	West Midlands Travel	1100	G100EOG	Leyland	B49F	09/89
LX1517	LX2R11C15Z4R	West Midlands Travel	1101	G101EOG	Leyland	B49F	09/89
LX1518	LX2R11C15Z4R	West Midlands Travel	1102	G102EOG	Leyland	B49F	09/89
LX1519	LX2R11C15Z4S	London Buses	LX4	G74UYV	Leyland	B49F	09/89
		London United	LX4	G74UYV	Leyland	B49F	11/94
LX1520	LX2R11C15Z4S	London Buses	LX5	G75UYV	Leyland	B49F	09/89
		London United	LX5	G75UYV	Leyland	B49F	11/94
LX1521	LX2R11C15Z4S	London Buses	LX6	G76UYV	Leyland	B49F	09/89
		London United	LX6	G76UYV	Leyland	B49F	11/94
LX1522	LX2R11C15Z4S	London Buses	LX7	G77UYV	Leyland	B49F	09/89
		London United	LX7	G77UYV	Leyland	B49F	11/94
LX1523	LX2R11C15Z4S	London Buses	LX8	G78UYV	Leyland	B49F	09/89
		London United	LX8	G78UYV	Leyland	B49F	11/94
LX1524	LX2R11C15Z4S	Boro'line, Maidstone	807	G40VME	Leyland	B49F	09/89
		Kentish Bus	407	G40VME	Leyland	B49F	02/92
LX1525	LX2R11C15Z4S	Boro'line, Maidstone	808	G41VME	Leyland	B49F	09/89
		Kentish Bus	408	G41VME	Leyland	B49F	02/92
LX1526	LX2R11C15Z4R	West Midlands Travel	1103	G103EOG	Leyland	B49F	09/89

G45VME was new to Boro'line of Maidstone and is seen in Crossharbour after being sold to Kentish Bus where it was used on the Docklands service D9. Interestingly, the vehicle has since moved to Maidstone & District returning, almost, to 'home'. *Tony Wilson*

LX1527	LX2R11C15Z4R	West Midlands Travel	1104	G104EOG	Leyland	B49F	09/89
LX1528	LX2R11C15Z4R	West Midlands Travel	1105	G105EOG	Leyland	B49F	09/89
LX1529	LX2R11C15Z4R	West Midlands Travel	1106	G106EOG	Leyland	B49F	09/89
LX1530	LX2R11C15Z4R	West Midlands Travel	1107	G107EOG	Leyland	B49F	09/89
LX1531	LX2R11C15Z4R	West Midlands Travel	1108	G108EOG	Leyland	B49F	09/89
LX1532	LX2R11C15Z4R	West Midlands Travel	1109	G109EOG	Leyland	B49F	09/89
LX1533	LX2R11C15Z4R	West Midlands Travel	1110	G110EOG	Leyland	B49F	09/89
LX1534	LX2R11C15Z4R	West Midlands Travel	1111	G111EOG	Leyland	B49F	09/89
LX1535	LX2R11C15Z4R	West Midlands Travel	1112	G112EOG	Leyland	B49F	09/89
LX1536	LX2R11C15Z4R	West Midlands Travel	1113	G113EOG	Leyland	B49F	09/89
LX1537	LX2R11C15Z4R	Halton	26	G222DKA	Leyland	B51F	10/89
LX1538	LX2R11C15Z4S	Boro'line, Maidstone	809	G42VME	Leyland	B49F	10/89
		Kentish Bus	409	G42VME	Leyland	B49F	02/92
LX1539	LX2R11C15Z4S	Boro'line, Maidstone	810	G43VME	Leyland	B49F	10/89
		Kentish Bus	410	G43VME	Leyland	B49F	02/92
LX1540	LX2R11C15Z4S	Boro'line, Maidstone	811	G44VME	Leyland	B49F	10/89
		Kentish Bus	411	G44VME	Leyland	B49F	02/92
LX1541	LX2R11C15Z4S	Boro'line, Maidstone	812	G45VME	Leyland	B49F	10/89
		Kentish Bus	412	G45VME	Leyland	B49F	02/92
		Maidstone & District	3050	G45VME	Leyland	B49F	09/95
LX1542	LX2R11C15Z4R	West Midlands Travel	1114	G114EOG	Leyland	B49F	09/89
LX1543	LX2R11C15Z4R	West Midlands Travel	1115	G115EOG	Leyland	B49F	09/89
LX1544	LX2R11C15Z4R	West Midlands Travel	1116	G116EOG	Leyland	B49F	09/89
LX1545	LX2R11C15Z4R	West Midlands Travel	1117	G117EOG	Leyland	B49F	09/89
LX1546	LX2R11C15Z4R	West Midlands Travel	1118	G118EOG	Leyland	B49F	09/89
LX1547	LX2R11C15Z4R	West Midlands Travel	1119	G119EOG	Leyland	B49F	09/89
LX1548	LX2R11C15Z4R	West Midlands Travel	1120	G120EOG	Leyland	B49F	09/89
LX1549	LX2R11C15Z4R	West Midlands Travel	1121	G121EOG	Leyland	B49F	09/89
LX1550	LX2R11C15Z4R	West Midlands Travel	1122	G122EOG	Leyland	B49F	09/89

United took delivery of five Lynx in 1990 and operated them in comparison trials with the Optare Delta. Shown here is LX1559 which entered service as United's 5005, G512EAJ. *David Cole*

LX1551	LX2R11C15Z4R	West Midlands Travel	1123	G123EOG	Leyland	B49F	09/89
LX1552	LX2R11C15Z4R	West Midlands Travel	1124	G124EOG	Leyland	B49F	09/89
LX1553	LX2R11C15Z4R	West Midlands Travel	1125	G125EOG	Leyland	B49F	09/89
LX1554	LX2R11C15Z4R	West Midlands Travel	1126	G126EOG	Leyland	B49F	10/89
LX1555	LX2R11C15Z4R	West Midlands Travel	1127	G127EOG	Leyland	B49F	10/89
LX1556	LX2R11C15Z4R	West Midlands Travel	1128	G128EOG	Leyland	B49F	10/89
LX1557	LX2R11C15Z4R	West Midlands Travel	1129	G129EOG	Leyland	B49F	10/89
LX1558	LX2R11C15Z4R	West Midlands Travel	1130	G130EOG	Leyland	B49F	10/89
LX1559	LX2R11C15Z4S	United	5005	G512EAJ	Leyland	B49F	01/90
LX1560	LX2R11C15Z4S	United	5001	G508EAJ	Leyland	B49F	01/90
LX1561	LX2R11C15Z4S	United	5002	G509EAJ	Leyland	B49F	01/90
LX1562	LX2R11C15Z4S	United	5003	G510EAJ	Leyland	B49F	01/90
LX1563	LX2R11C15Z4S	United	5004	G511EAJ	Leyland	B49F	01/90
LX1564	LX2R11C15Z4R	West Midlands Travel	1131	G131EOG	Leyland	B49F	10/89
LX1565	LX2R11C15Z4R	West Midlands Travel	1132	G132EOG	Leyland	B49F	10/89
LX1566	LX2R11C15Z4R	West Midlands Travel	1133	G133EOG	Leyland	B49F	10/89
LX1567	LX2R11C15Z4R	West Midlands Travel	1134	G134EOG	Leyland	B49F	10/89
LX1568	LX2R11C15Z4R	West Midlands Travel	1135	G135EOG	Leyland	B49F	10/89
LX1569	LX2R11C15Z4R	West Midlands Travel	1136	G136EOG	Leyland	B49F	10/89
LX1570	LX2R11C15Z4R	West Midlands Travel	1137	G137EOG	Leyland	B49F	10/89
LX1571	LX2R11C15Z4R	West Midlands Travel	1138	G138EOG	Leyland	B49F	10/89
LX1572	LX2R11C15Z4R	West Midlands Travel	1139	G139EOG	Leyland	B49F	10/89
LX1573	LX2R11C15Z4R	West Midlands Travel	1140	G140EOG	Leyland	B49F	10/89
LX1574	LX2R11C15Z4S	Sovereign Bus	201	G201URO	Leyland	B49F	10/89
LX1575	LX2R11C15Z4S	Sovereign Bus	202	G202URO	Leyland	B49F	10/89
LX1576	LX2R11C15Z4S	Sovereign Bus	203	G203URO	Leyland	B49F	10/89
LX1577	LX2R11C15Z4S	Sovereign Bus	204	G204URO	Leyland	B49F	10/89

LX1578	LX2R11C15Z4S	Sovereign Bus	205	G205URO	Leyland	B49F	10/89
LX1579	LX2R11C15Z4S	Sovereign Bus	206	G206URO	Leyland	B49F	10/89
LX1580	LX2R11C15Z4S	Sovereign Bus	207	G207URO	Leyland	B49F	10/89
LX1581	LX2R11C15Z4R	West Midlands Travel	1141	G141EOG	Leyland	B49F	10/89
LX1582	LX2R11C15Z4R	West Midlands Travel	1142	G142EOG	Leyland	B49F	10/89
LX1583	LX2R11C15Z4R	West Midlands Travel	1143	G143EOG	Leyland	B49F	10/89
LX1584	LX2R11C15Z4R	West Midlands Travel	1144	G144EOG	Leyland	B49F	10/89
LX1585	LX2R11C15Z4R	West Midlands Travel	1145	G145EOG	Leyland	B49F	10/89
LX1586	LX2R11C15Z4R	West Midlands Travel	1146	G146EOG	Leyland	B49F	10/89
LX1587	LX2R11C15Z4R	West Midlands Travel	1147	G147EOG	Leyland	B49F	10/89
LX1588	LX2R11C15Z4R	West Midlands Travel	1148	G148EOG	Leyland	B49F	10/89
LX1589	LX2R11C15Z4R	West Midlands Travel	1149	G149EOG	Leyland	B49F	10/89
LX1590	LX2R11C15Z4R	West Midlands Travel	1150	G150EOG	Leyland	B49F	10/89
LX1591	LX2R11C15Z4R	West Midlands Travel	1151	G151EOG	Leyland	B49F	10/89
LX1592	LX2R11C15Z4R	West Midlands Travel	1152	G152EOG	Leyland	B49F	10/89
LX1593	LX2R11C15Z4R	West Midlands Travel	1153	G153EOG	Leyland	B49F	10/89
LX1594	LX2R11C15Z4R	West Midlands Travel	1154	G154EOG	Leyland	B49F	10/89
LX1595	LX2R11C15Z4R	West Midlands Travel	1155	G155EOG	Leyland	B49F	10/89
LX1596	LX2R11C15Z4R	West Midlands Travel	1156	G156EOG	Leyland	B49F	10/89
LX1597	LX2R11C15Z4R	West Midlands Travel	1157	G157EOG	Leyland	B49F	10/89
LX1598	LX2R11C15Z4S	West Riding	329	G329NUM	Leyland	B49F	04/90
LX1599	LX2R11C15Z4S	Yorkshire	318	G317NNW	Leyland	B49F	03/90
LX1600	LX2R11C15Z4S	Yorkshire	319	G319NNW	Leyland	B49F	03/90
LX1601	LX2R11C15Z4S	Yorkshire	320	G324NNW	Leyland	B49F	03/90
LX1602	LX2R11C15Z4S	Yorkshire	321	G321NNW	Leyland	B49F	03/90
LX1603	LX2R11C15Z4S	Yorkshire	322	G322NNW	Leyland	B49F	03/90
LX1604	LX2R11C15Z4S	Yorkshire	328	G109OUG	Leyland	B49F	04/90
LX1605	LX2R11G15Z4S	London & Country	311	G311DPA	Leyland	B49F	01/90
		London Links	311	G311DPA	Leyland	B49F	01/95
LX1606	LX2R11C15Z4R	Preston Bus	14	G214KRN	Leyland	DP45F	11/89
LX1607	LX2R11C15Z4R	Preston Bus	15	G215KRN	Leyland	DP45F	11/89
LX1608	LX2R11C15Z4R	Preston Bus	16	G216KRN	Leyland	DP45F	11/89
LX1609	LX2R11C15Z4R	Preston Bus	17	G217KRN	Leyland	DP45F	11/89
LX1610	LX2R11C15Z4R	Preston Bus	18	G218KRN	Leyland	DP45F	11/89
LX1611	LX2R11C15Z4R	West Midlands Travel	1158	G158EOG	Leyland	B49F	11/89
LX1612	LX2R11C15Z4R	West Midlands Travel	1159	G159EOG	Leyland	B49F	11/89
LX1613	LX2R11C15Z4R	West Midlands Travel	1160	G160EOG	Leyland	B49F	11/89
LX1614	LX2R11C15Z4R	West Midlands Travel	1161	G161EOG	Leyland	B49F	11/89
LX1615	LX2R11C15Z4R	West Midlands Travel	1162	G162EOG	Leyland	B49F	11/89
LX1616	LX2R11C15Z4R	West Midlands Travel	1163	G163EOG	Leyland	B49F	11/89
LX1617	LX2R11C15Z4R	West Midlands Travel	1164	G164EOG	Leyland	B49F	11/89
LX1618	LX2R11C15Z4R	West Midlands Travel	1165	G165EOG	Leyland	B49F	11/89
LX1619	LX2R11C15Z4R	West Midlands Travel	1166	G166EOG	Leyland	B49F	11/89
LX1620	LX2R11C15Z4R	West Midlands Travel	1167	G167EOG	Leyland	B49F	11/89
LX1621	LX2R11C15Z4R	West Midlands Travel	1168	G168EOG	Leyland	B49F	11/89
LX1622	LX2R11C15Z4R	Colchester	38	G38YHJ	Leyland	B49F	11/89
		Crosville Wales	SLC38	G38YHJ	Leyland	B49F	05/94
LX1623	LX2R11C15Z4R	Colchester	39	G39YHJ	Leyland	B49F	11/89
		Crosville Wales	SLC39	G39YHJ	Leyland	B49F	05/94
LX1624	LX2R11C15Z4R	Colchester	40	G40YHJ	Leyland	B49F	11/89
		Crosville Wales	SLC40	G40YHJ	Leyland	B49F	08/94
LX1625	LX2R11C15Z4R	West Midlands Travel	1169	G169EOG	Leyland	B49F	11/89
LX1626	LX2R11C15Z4R	West Midlands Travel	1170	G170EOG	Leyland	B49F	11/89
LX1627	LX2R11C15Z4R	West Midlands Travel	1171	G171EOG	Leyland	B49F	11/89
LX1628	LX2R11C15Z4R	West Midlands Travel	1172	G172EOG	Leyland	B49F	11/89
LX1629	LX2R11C15Z4S	Yorkshire	323	G108OUG	Leyland	B49F	04/90
LX1630	LX2R11C15Z4S	Yorkshire	324	G324NUM	Leyland	B49F	04/90

Preston Bus operate fifteen Lynx, all bar two with high-back seating. Photographed leaving the pagoda-shaped bus station for Lea is 17, G217KRN one of five delivered in 1989. *Daniel Hill*

Crosville Wales have been acquiring Lynx from several other British Bus fleets during 1994 and 1995. Notable were the large number from Colchester that displaced National Greenways. Seen in Chester shortly after its acquisition is SLC40, G40YHJ. *Malc McDonald*

LX1631	LX2R11C15Z4S	Yorkshire	325	G110OUG	Leyland	B49F	04/90
LX1632	LX2R11C15Z4S	Yorkshire	326	G326NUM	Leyland	B49F	04/90
LX1633	LX2R11C15Z4S	Yorkshire	327	G327NUM	Leyland	B49F	04/90
LX1634	LX2R11C15Z4S	West Riding	332	G332NUM	Leyland	B49F	04/90
LX1635	LX2R11C15Z4R	West Midlands Travel	1173	G173EOG	Leyland	B49F	11/89
LX1636	LX2R11C15Z4R	West Midlands Travel	1174	G174EOG	Leyland	B49F	11/89
LX1637	LX2R11C15Z4R	West Midlands Travel	1175	G175EOG	Leyland	B49F	11/89
LX1638	LX2R11C15Z4R	West Midlands Travel	1176	G176EOG	Leyland	B49F	11/89
LX1639	LX2R11C15Z4R	West Midlands Travel	1177	G177EOG	Leyland	B49F	11/89
LX1640	LX2R11C15Z4R	West Midlands Travel	1178	G178EOG	Leyland	B49F	11/89
LX1641	LX2R11C15Z4R	West Midlands Travel	1179	G179EOG	Leyland	B49F	11/89
LX1642	LX2R11C15Z4R	West Midlands Travel	1180	G180EOG	Leyland	B49F	11/89
LX1643	LX2R11C15Z4R	West Midlands Travel	1181	G181EOG	Leyland	B49F	11/89
LX1644	LX2R11C15Z4R	West Midlands Travel	1182	G182EOG	Leyland	B49F	12/89
LX1645	LX2R11C15Z4R	City of Nottingham	745	G745PNN	Leyland	B50F	12/89
LX1646	LX2R11C15Z4R	City of Nottingham	746	G746PNN	Leyland	B50F	12/89
LX1647	LX2R11C15Z4R	City of Nottingham	747	G747PNN	Leyland	B50F	12/89
LX1648	LX2R11C15Z4R	City of Nottingham	748	G748PNN	Leyland	B50F	12/89
LX1649	LX2R11C15Z4R	City of Nottingham	749	G749PNN	Leyland	B50F	12/89
LX1650	LX2R11C15Z4R	West Midlands Travel	1183	G183EOG	Leyland	B49F	12/89
LX1651	LX2R11C15Z4R	West Midlands Travel	1184	G184EOG	Leyland	B49F	12/89
LX1652	LX2R11C15Z4R	West Midlands Travel	1185	G185EOG	Leyland	B49F	12/89
LX1653	LX2R11C15Z4R	West Midlands Travel	1186	G186EOG	Leyland	B49F	12/89
LX1654	LX2R11C15Z4R	West Midlands Travel	1187	G187EOG	Leyland	B49F	12/89
LX1655	LX2R11C15Z4R	West Midlands Travel	1188	G188EOG	Leyland	B49F	01/90
LX1656	LX2R11C15Z4R	West Midlands Travel	1189	G189EOG	Leyland	B49F	01/90
LX1657	LX2R11C15Z4R	West Midlands Travel	1190	G190EOG	Leyland	B49F	01/90
LX1658	LX2R11C15Z4R	West Midlands Travel	1191	G191EOG	Leyland	B49F	01/90
LX1659	LX2R11C15Z4R	West Midlands Travel	1192	G192EOG	Leyland	B49F	01/90
LX1660	LX2R11C15Z4R	West Midlands Travel	1193	G193EOG	Leyland	B49F	01/90
LX1661	LX2R11C15Z4R	West Midlands Travel	1194	G194EOG	Leyland	B49F	01/90
LX1662	LX2R11C15Z4R	West Midlands Travel	1195	G195EOG	Leyland	B49F	01/90
LX1663	LX2R11C15Z4R	West Midlands Travel	1196	G196EOG	Leyland	B49F	01/90
LX1664	LX2R11C15Z4R	West Midlands Travel	1197	G197EOG	Leyland	B49F	01/90
LX1665	LX2R11C15Z4R	West Midlands Travel	1198	G198EOG	Leyland	B49F	01/90
LX1666	LX2R11C15Z4R	West Midlands Travel	1199	G199EOG	Leyland	B49F	01/90
LX1667	LX2R11C15Z4R	West Midlands Travel	1200	G200EOG	Leyland	B49F	01/90
LX1668	LX2R11C15Z4R	West Midlands Travel	1201	G201EOG	Leyland	B49F	01/90
LX1669	LX2R11C15Z4R	West Midlands Travel	1202	G202EOG	Leyland	B49F	01/90
LX1670	LX2R11C15Z4R	West Midlands Travel	1203	G203EOG	Leyland	B49F	01/90
LX1671	LX2R11C15Z4R	West Midlands Travel	1204	G204EOG	Leyland	B49F	01/90
LX1672	LX2R11C15Z4R	West Midlands Travel	1205	G205EOG	Leyland	B49F	01/90
LX1673	LX2R11C15Z4R	West Midlands Travel	1206	G206EOG	Leyland	B49F	01/90
LX1674	LX2R11G15Z4S	London & Country	312	G312DPA	Leyland	B49F	01/90
		LondonLinks	312	G312DPA	Leyland	B49F	01/95
LX1675	LX2R11G15Z4S	London & Country	313	G313DPA	Leyland	B49F	01/90
		LondonLinks	313	G313DPA	Leyland	B49F	01/95
LX1676	LX2R11G15Z4S	London & Country	314	G314DPA	Leyland	B49F	01/90
		LondonLinks	314	G314DPA	Leyland	B49F	01/95
LX1677	LX2R11G15Z4S	London & Country	315	G315DPA	Leyland	B49F	01/90
		LondonLinks	315	G315DPA	Leyland	B49F	01/95
LX1678	LX2R11G15Z4S	London & Country	316	G316DPA	Leyland	B49F	01/90
		LondonLinks	316	G316DPA	Leyland	B49F	01/95
LX1679	LX2R11C15Z4R	Midland Red West	1101	G101HNP	Leyland	B49F	02/90
LX1680	LX2R11C15Z4R	West Midlands Travel	1207	G207EOG	Leyland	B49F	01/90
LX1681	LX2R11C15Z4R	West Midlands Travel	1208	G208EOG	Leyland	B49F	01/90
LX1682	LX2R11C15Z4R	West Midlands Travel	1209	G209EOG	Leyland	B49F	01/90

Five Gardner-engined Lynx joined London & Country in 1990 in comparative trials with Dennis Falcons. Seen in this attractive livery is 315, G315DPA. *David Cole*

The latest livery for West Midlands Travel is shown on LX1806. Numbered 1252, G252EOG, the delivery of 250 buses does not run in sequence, being interspersed with fifteen built with high-back seating. *Daniel Hill*

Midland Red West's major investment in the Lynx consisted of fifty examples which were all allocated to the Digbeth depot in Birmingham. From here they operated on a network of Commercial and tendered services in Birmingham and the Black Country. Subsequently the type has been re-allocated to other depots. Here is shown LX 1720 as G115HNP, at Redditch. *Ralph Stevens*

LX1683	LX2R11C15Z4R	West Midlands Travel	1210	G210EOG	Leyland		B49F	01/90
LX1684	LX2R11C15Z4R	West Midlands Travel	1211	G211EOG	Leyland		B49F	01/90
LX1685	LX2R11C15Z4R	West Midlands Travel	1212	G212EOG	Leyland		B49F	01/90
LX1686	LX2R11C15Z4R	West Midlands Travel	1213	G213EOG	Leyland		B49F	01/90
LX1687	LX2R11C15Z4R	Action, Canberra, AUS	730	BUS730	Leyland		B49F	06/91
		Hadfields, Sydney, AUS		MO8499	Leyland		B47F	??
LX1688	LX2R11C15Z4R	Action, Canberra, AUS	731	BUS731	Leyland		B49F	06/91
		Hadfields, Sydney, AUS		MO8501	Leyland		B47F	??
LX1689	LX2R11C15Z4R	West Midlands Travel	1255	G255EOG	Leyland		DP47F	04/90
LX1690	LX2R11G15Z4S	Boro'line, Maidstone	813	H813EKJ	Leyland		B49F	02/91
		Kentish Bus	413	H813EKJ	Leyland		B49F	02/92
LX1691	LX2R11G15Z4S	Boro'line, Maidstone	814	H814EKJ	Leyland		B49F	02/91
		Kentish Bus	414	H814EKJ	Leyland		B49F	02/92
LX1692	LX2R11G15Z4S	Boro'line, Maidstone	815	H815EKJ	Leyland		B49F	02/91
		Kentish Bus	415	H815EKJ	Leyland		B49F	02/92
LX1693	LX2R11G15Z4S	Boro'line, Maidstone	816	H816EKJ	Leyland		B49F	02/91
		Kentish Bus	416	H816EKJ	Leyland		B49F	02/92
LX1694	LX2R11G15Z4R	Colchester	28	H28MJN	Leyland		B49F	02/91
		Crosville Wales	SLG28	H28MJN	Leyland		B49F	10/94
LX1695	LX2R11G15Z4R	Colchester	29	H29MJN	Leyland		B49F	02/91
		Crosville Wales	SLG29	H29MJN	Leyland		B49F	10/94
LX1696	LX2R11C15Z4R	Midland Red West	1102	G102HNP	Leyland		B49F	02/90
LX1697	LX2R11C15Z4R	Midland Red West	1103	G103HNP	Leyland		B49F	02/90
LX1698	LX2R11C15Z4R	Midland Red West	1104	G104HNP	Leyland		B49F	02/90
LX1699	LX2R11C15Z4R	Midland Red West	1105	G105HNP	Leyland		B49F	02/90
LX1700	LX2R11C15Z4R	West Midlands Travel	1214	G214EOG	Leyland		B49F	01/90

LX1701	LX2R11C15Z4R	West Midlands Travel	1215	G215EOG	Leyland	B49F	01/90
LX1702	LX2R11C15Z4R	West Midlands Travel	1216	G216EOG	Leyland	B49F	01/90
LX1703	LX2R11C15Z4R	West Midlands Travel	1217	G217EOG	Leyland	B49F	01/90
LX1704	LX2R11C15Z4R	Midland Red West	1106	G106HNP	Leyland	B49F	04/90
LX1705	LX2R11C15Z4R	Midland Red West	1107	G107HNP	Leyland	B49F	04/90
LX1706	LX2R11C15Z4R	Midland Red West	1108	G108HNP	Leyland	B49F	04/90
LX1707	LX2R11C15Z4R	Midland Red West	1109	G109HNP	Leyland	B49F	04/90
LX1708	LX2R11C15Z4R	Midland Red West	1110	G110HNP	Leyland	B49F	04/90
LX1709	LX2R11C15Z4R	Midland Red West	1111	G111HNP	Leyland	B49F	04/90
LX1710	LX2R11C15Z4R	Midland Red West	1112	G112HNP	Leyland	B49F	04/90
LX1711	LX2R11C15Z4R	Midland Red West	1113	G113HNP	Leyland	B49F	04/90
LX1712	LX2R11C15Z4R	West Midlands Travel	1218	G218EOG	Leyland	B49F	01/90
LX1713	LX2R11C15Z4R	West Midlands Travel	1219	G219EOG	Leyland	B49F	04/90
LX1714	LX2R11C15Z4R	West Midlands Travel	1220	G220EOG	Leyland	B49F	02/90
LX1715	LX2R11C15Z4R	West Midlands Travel	1221	G221EOG	Leyland	B49F	04/90
LX1716	LX2R11C15Z4R	West Midlands Travel	1222	G222EOG	Leyland	B49F	04/90
LX1717	LX2R11C15Z4R	West Midlands Travel	1223	G223EOG	Leyland	B49F	02/90
LX1718	LX2R11C15Z4R	West Midlands Travel	1224	G224EOG	Leyland	B49F	02/90
LX1719	LX2R11C15Z4R	Midland Red West	1114	G114HNP	Leyland	B49F	04/90
LX1720	LX2R11C15Z4R	Midland Red West	1115	G115HNP	Leyland	B49F	04/90
LX1721	LX2R11C15Z4R	Midland Red West	1116	G116HNP	Leyland	B49F	04/90
LX1722	LX2R11C15Z4R	Midland Red West	1117	G117HNP	Leyland	B49F	04/90
LX1723	LX2R11C15Z4R	Midland Red West	1118	G118HNP	Leyland	B49F	04/90
LX1724	LX2R11C15Z4R	Midland Red West	1119	G119HNP	Leyland	B49F	04/90
LX1725	LX2R11C15Z4R	Midland Red West	1120	G120HNP	Leyland	B49F	04/90
LX1726	LX2R11C15Z4R	West Midlands Travel	1225	G225EOG	Leyland	B49F	04/90
LX1727	LX2R11C15Z4R	West Midlands Travel	1226	G226EOG	Leyland	B49F	04/90
LX1728	LX2R11C15Z4R	West Midlands Travel	1227	G227EOG	Leyland	B49F	04/90
LX1729	LX2R11C15Z4R	West Midlands Travel	1228	G228EOG	Leyland	B49F	04/90
LX1730	LX2R11C15Z4R	West Midlands Travel	1229	G229EOG	Leyland	B49F	04/90
LX1731	LX2R11C15Z4R	West Midlands Travel	1256	G256EOG	Leyland	DP47F	04/90
LX1732	LX2R11C15Z4R	West Midlands Travel	1257	G257EOG	Leyland	DP47F	04/90
LX1733	LX2R11C15Z4R	West Midlands Travel	1258	G258EOG	Leyland	DP47F	04/90
LX1734	LX2R11C15Z4R	West Midlands Travel	1259	G259EOG	Leyland	DP47F	04/90
LX1735	LX2R11C15Z4R	Midland Red West	1121	G121HNP	Leyland	B49F	04/90
LX1736	LX2R11C15Z4R	Midland Red West	1122	G122HNP	Leyland	B49F	04/90
LX1737	LX2R11C15Z4R	Midland Red West	1123	G123HNP	Leyland	B49F	04/90
LX1738	LX2R11C15Z4R	Midland Red West	1124	G124HNP	Leyland	B49F	04/90
LX1739	LX2R11C15Z4R	Midland Red West	1125	G125HNP	Leyland	B49F	04/90
LX1740	LX2R11C15Z4R	Midland Red West	1126	G126HNP	Leyland	B49F	04/90
LX1741	LX2R11C15Z4R	Midland Red West	1127	G127HNP	Leyland	B49F	04/90
LX1742	LX2R11C15Z4R	West Midlands Travel	1260	G260EOG	Leyland	DP47F	04/90
LX1743	LX2R11C15Z4R	West Midlands Travel	1261	G261EOG	Leyland	DP47F	04/90
LX1744	LX2R11C15Z4R	West Midlands Travel	1262	G262EOG	Leyland	DP47F	04/90
LX1745	LX2R11C15Z4R	Midland Red West	1128	G128HNP	Leyland	B49F	04/90
LX1746	LX2R11C15Z4R	Midland Red West	1129	G129HNP	Leyland	B49F	04/90
LX1747	LX2R11C15Z4R	Midland Red West	1130	G130HNP	Leyland	B49F	04/90
LX1748	LX2R11C15Z4R	Midland Red West	1131	G131HNP	Leyland	B49F	04/90
LX1749	LX2R11C15Z4R	Midland Red West	1132	G132HNP	Leyland	B49F	04/90
LX1750	LX2R11C15Z4R	Midland Red West	1133	G133HNP	Leyland	B49F	04/90
LX1751	LX2R11C15Z4R	West Midlands Travel	1263	G263EOG	Leyland	DP47F	04/90
LX1752	LX2R11C15Z4R	West Midlands Travel	1264	G264EOG	Leyland	DP47F	04/90
LX1753	LX2R11C15Z4R	West Midlands Travel	1265	G265EOG	Leyland	DP47F	04/90
LX1754	LX2R11C15Z4R	West Midlands Travel	1266	G266EOG	Leyland	DP47F	04/90
LX1755	LX2R11C15Z4R	West Midlands Travel	1267	G267EOG	Leyland	DP47F	04/90
LX1756	LX2R11C15Z4R	West Midlands Travel	1268	G268EOG	Leyland	DP47F	04/90
LX1757	LX2R11C15Z4R	West Midlands Travel	1269	G269EOG	Leyland	DP47F	04/90

The Leyland Lynx

County's TownLink marketing name is seen on LX1777 their LX253, H253GEV. This operator chose the flat floor and stepped rear in preference to the ramped alternative. *Tony Wilson*

LX1758	LX2R11C15Z4S	County	LX251	H251GEV	Leyland	B49F	07/90
LX1759	LX2R11C15Z4S	County	LX252	H252GEV	Leyland	B49F	07/90
LX1760	LX2R11C15Z4R	West Midlands Travel	1230	G230EOG	Leyland	B49F	04/90
LX1761	LX2R11C15Z4R	West Midlands Travel	1231	G231EOG	Leyland	B49F	04/90
LX1762	LX2R11C15Z4R	Midland Red West	1134	G134HNP	Leyland	B49F	04/90
LX1763	LX2R11C15Z4R	Midland Red West	1135	G135HNP	Leyland	B49F	04/90
LX1764	LX2R11C15Z4R	Midland Red West	1136	G136HNP	Leyland	B49F	04/90
LX1765	LX2R11C15Z4R	Midland Red West	1137	G137HNP	Leyland	B49F	04/90
LX1766	LX2R11C15Z4R	Midland Red West	1138	G138HNP	Leyland	B49F	04/90
LX1767	LX2R11C15Z4R	Midland Red West	1139	G139HNP	Leyland	B49F	04/90
LX1768	LX2R11C15Z4R	Midland Red West	1140	G140HNP	Leyland	B49F	04/90
LX1769	LX2R11C15Z4S	West Riding	330	G330NUM	Leyland	B49F	04/90
LX1770	LX2R11C15Z4S	West Riding	330	G330NUM	Leyland	B49F	04/90
LX1771	LX2R11C15Z4R	Midland Red West	1141	G141HNP	Leyland	B49F	04/90
LX1772	LX2R11C15Z4R	Midland Red West	1142	G142HNP	Leyland	B49F	04/90
LX1773	LX2R11C15Z4R	Midland Red West	1143	G143HNP	Leyland	B49F	04/90
LX1774	LX2R11C15Z4R	Midland Red West	1144	G144HNP	Leyland	B49F	04/90
LX1775	LX2R11C15Z4R	Midland Red West	1145	G145HNP	Leyland	B49F	04/90
LX1776	LX2R11C15Z4R	Midland Red West	1146	G146HNP	Leyland	B49F	04/90
LX1777	LX2R11C15Z4S	County	LX253	H253GEV	Leyland	B49F	07/90
LX1778	LX2R11C15Z4R	Midland Red West	1147	G147HNP	Leyland	B49F	04/90
LX1779	LX2R11C15Z4R	Midland Red West	1148	G148HNP	Leyland	B49F	04/90
LX1780	LX2R11C15Z4R	Midland Red West	1149	G149HNP	Leyland	B49F	04/90
LX1781	LX2R11C15Z4R	Midland Red West	1150	G150HNP	Leyland	B49F	04/90
LX1782	LX2R11C15Z4R	West Midlands Travel	1232	G232EOG	Leyland	B49F	04/90
LX1783	LX2R11C15Z4R	West Midlands Travel	1233	G233EOG	Leyland	B49F	04/90
LX1784	LX2R11C15Z4R	West Midlands Travel	1234	G234EOG	Leyland	B49F	04/90
LX1785	LX2R11C15Z4R	West Midlands Travel	1235	G235EOG	Leyland	B49F	04/90

LX1786	LX2R11C15Z4R	West Midlands Travel	1236	G236EOG	Leyland	B49F	04/90
LX1787	LX2R11C15Z4R	West Midlands Travel	1237	G237EOG	Leyland	B49F	04/90
LX1788	LX2R11C15Z4R	West Midlands Travel	1238	G238EOG	Leyland	B49F	04/90
LX1789	LX2R11C15Z4R	West Midlands Travel	1239	G239EOG	Leyland	B49F	04/90
LX1790	LX2R11C15Z4R	West Midlands Travel	1240	G240EOG	Leyland	B49F	04/90
LX1791	LX2R11C15Z4R	West Midlands Travel	1241	G241EOG	Leyland	B49F	04/90
LX1792	LX2R11C15Z4R	West Midlands Travel	1242	G242EOG	Leyland	B49F	04/90
LX1793	LX2R11C15Z4R	West Midlands Travel	1243	G243EOG	Leyland	B49F	04/90
LX1794	LX2R11C15Z4R	West Midlands Travel	1244	G244EOG	Leyland	B49F	04/90
LX1795	LX2R11C15Z4R	West Midlands Travel	1245	G245EOG	Leyland	B49F	05/90
LX1796	LX2R11C15Z4S	County	LX254	H254GEV	Leyland	B49F	07/90
LX1797	LX2R11C15Z4S	County	LX255	H255GEV	Leyland	B49F	07/90
LX1798	LX2R11C15Z4S	County	LX256	H256GEV	Leyland	B49F	07/90
LX1799	LX2R11C15Z4S	County	LX257	H257GEV	Leyland	B49F	07/90
LX1800	LX2R11C15Z4R	West Midlands Travel	1246	G246EOG	Leyland	B49F	05/90
LX1801	LX2R11C15Z4R	West Midlands Travel	1247	G247EOG	Leyland	B49F	05/90
LX1802	LX2R11C15Z4R	West Midlands Travel	1248	G248EOG	Leyland	B49F	05/90
LX1803	LX2R11C15Z4R	West Midlands Travel	1249	G249EOG	Leyland	B49F	05/90
LX1804	LX2R11C15Z4R	West Midlands Travel	1250	G250EOG	Leyland	B49F	05/90
LX1805	LX2R11C15Z4R	West Midlands Travel	1251	G251EOG	Leyland	B49F	05/90
LX1806	LX2R11C15Z4R	West Midlands Travel	1252	G252EOG	Leyland	B49F	05/90
LX1807	LX2R11C15Z4R	West Midlands Travel	1253	G253EOG	Leyland	B49F	05/90
LX1808	LX2R11C15Z4R	West Midlands Travel	1254	G254EOG	Leyland	B49F	05/90
LX1809	LX2R11C15Z4R	West Midlands Travel	1270	G270EOG	Leyland	B49F	05/90
LX1810	LX2R11C15Z4R	West Midlands Travel	1271	G271EOG	Leyland	B49F	05/90
LX1811	LX2R11C15Z4R	West Midlands Travel	1272	G272EOG	Leyland	B49F	05/90
LX1812	LX2R11C15Z4R	West Midlands Travel	1273	G273EOG	Leyland	B49F	05/90
LX1813	LX2R11C15Z4R	West Midlands Travel	1274	G274EOG	Leyland	B49F	05/90
LX1814	LX2R11C15Z4R	West Midlands Travel	1275	G275EOG	Leyland	B49F	05/90
LX1815	LX2R11C15Z4R	West Midlands Travel	1276	G276EOG	Leyland	B49F	05/90
LX1816	LX2R11C15Z4R	West Midlands Travel	1277	G277EOG	Leyland	B49F	05/90
LX1817	LX2R11C15Z4R	West Midlands Travel	1278	G278EOG	Leyland	B49F	05/90
LX1818	LX2R11C15Z4R	West Midlands Travel	1279	G279EOG	Leyland	B49F	05/90
LX1819	LX2R11C15Z4R	West Midlands Travel	1280	G280EOG	Leyland	B49F	06/90
LX1820	LX2R11C15Z4R	West Midlands Travel	1281	G281EOG	Leyland	B49F	06/90
LX1821	LX2R11C15Z4R	West Midlands Travel	1282	G282EOG	Leyland	B49F	06/90
LX1822	LX2R11C15Z4R	West Midlands Travel	1283	G283EOG	Leyland	B49F	06/90
LX1823	LX2R11C15Z4R	West Midlands Travel	1284	G284EOG	Leyland	B49F	06/90
LX1824	LX2R11C15Z4R	West Midlands Travel	1285	G285EOG	Leyland	B49F	06/90
LX1825	LX2R11C15Z4R	West Midlands Travel	1286	G286EOG	Leyland	B49F	06/90
LX1826	LX2R11C15Z4R	West Midlands Travel	1287	G287EOG	Leyland	B49F	06/90
LX1827	LX2R11C15Z4R	West Midlands Travel	1288	G288EOG	Leyland	B49F	06/90
LX1828	LX2R11C15Z4R	West Midlands Travel	1289	G289EOG	Leyland	B49F	06/90
LX1829	LX2R11C15Z4R	West Midlands Travel	1290	G290EOG	Leyland	B49F	06/90
LX1830	LX2R11C15Z4R	West Midlands Travel	1291	G291EOG	Leyland	B49F	06/90
LX1831	LX2R11C15Z4R	West Midlands Travel	1292	G292EOG	Leyland	B49F	06/90
LX1832	LX2R11C15Z4R	West Midlands Travel	1293	G293EOG	Leyland	B49F	06/90
LX1833	LX2R11C15Z4R	West Midlands Travel	1294	G294EOG	Leyland	B49F	06/90
LX1834	LX2R11C15Z4R	West Midlands Travel	1295	G295EOG	Leyland	B49F	06/90
LX1835	LX2R11C15Z4R	West Midlands Travel	1296	G296EOG	Leyland	B49F	06/90
LX1836	LX2R11C15Z4R	West Midlands Travel	1297	G297EOG	Leyland	B49F	06/90
LX1837	LX2R11C15Z4R	West Midlands Travel	1298	G298EOG	Leyland	B49F	06/90
LX1838	LX2R11C15Z4R	West Midlands Travel	1299	G299EOG	Leyland	B49F	06/90
LX1839	LX2R11C15Z4R	West Midlands Travel	1300	G300EOG	Leyland	B49F	06/90
LX1840	LX2R11C15Z4R	West Midlands Travel	1301	G301EOG	Leyland	B49F	06/90
LX1841	LX2R11C15Z4R	West Midlands Travel	1302	G302EOG	Leyland	B49F	06/90
LX1842	LX2R11C15Z4R	City Line	1633	H633YHT	Leyland	B49F	08/90

Ten Lynx with high-back seating were placed in service with PMT during 1990. At that time the PMT fleet were numbered in a simple numeric series which has been given prefixes similar to the Crosville system. Here SLC855, H855GRE is seen heading for Meir Square. *Cliff Beeton*

LX1843	LX2R11C15Z4R	West Midlands Travel	1303	G303EOG	Leyland	B49F	06/90
LX1844	LX2R11C15Z4R	West Midlands Travel	1304	G304EOG	Leyland	B49F	06/90
LX1845	LX2R11C15Z4R	West Midlands Travel	1305	G305EOG	Leyland	B49F	06/90
LX1846	LX2R11C15Z4R	West Midlands Travel	1306	G306EOG	Leyland	B49F	06/90
LX1847	LX2R11C15Z4R	West Midlands Travel	1307	G307EOG	Leyland	B49F	06/90
LX1848	LX2R11C15Z4R	Volvo Bus, Canberra, AUS		?	Leyland	B49F	01/91
		Hadfields, Sydney, AUS		MO8500	Leyland	B47F	00/91
LX1849	LX2R11C15Z4R	West Midlands Travel	1308	G308EOG	Leyland	B49F	06/90
LX1850	LX2R11C15Z4R	West Midlands Travel	1309	G309EOG	Leyland	B49F	06/90
LX1851	LX2R11C15Z4R	West Midlands Travel	1310	G310EOG	Leyland	B49F	06/90
LX1852	LX2R11C15Z4R	West Midlands Travel	1311	G311EOG	Leyland	B49F	06/90
LX1853	LX2R11C15Z4R	West Midlands Travel	1312	G312EOG	Leyland	B49F	06/90
LX1854	LX2R11C15Z4R	West Midlands Travel	1313	G313EOG	Leyland	B49F	06/90
LX1855	LX2R11C15Z4R	West Midlands Travel	1314	G314EOG	Leyland	B49F	07/90
LX1856	LX2R11C15Z4R	West Midlands Travel	1315	G315EOG	Leyland	B49F	07/90
LX1857	LX2R11C15Z4R	PMT	SLC851	H851GRE	Leyland	DP48F	08/90
LX1858	LX2R11C15Z4R	PMT	SLC852	H852GRE	Leyland	DP48F	08/90
LX1859	LX2R11C15Z4R	PMT	SLC853	H853GRE	Leyland	DP48F	08/90
LX1860	LX2R11C15Z4R	PMT	SLC854	H854GRE	Leyland	DP48F	08/90
LX1861	LX2R11C15Z4R	PMT	SLC855	H855GRE	Leyland	DP48F	08/90
LX1862	LX2R11C15Z4R	PMT	SLC856	H856GRE	Leyland	DP48F	08/90
LX1863	LX2R11C15Z4R	PMT	SLC857	H857GRE	Leyland	DP48F	08/90
LX1864	LX2R11C15Z4R	PMT	SLC858	H858GRE	Leyland	DP48F	08/90
LX1865	LX2R11C15Z4R	PMT	SLC859	H859GRE	Leyland	DP48F	08/90
LX1866	LX2R11C15Z4R	PMT	SLC860	H860GRE	Leyland	DP48F	08/90
LX1867	LX2R11C15Z4R	PMT	SLC861	H861GRE	Leyland	DP48F	08/90

Another FirstBus company to operate the Lynx is City Line of Bristol. A batch of 33 was delivered in 1989 followed in 1990 by a further 30. Seen with the original fleetname is 1652, H652YHT. The later style of fleetname can be seen in the 1996 FirstBus Bus Handbook. *David Cole*

LX1868	LX2R11C15Z4S	County	LX258	H258GEV	Leyland	B49F	08/90
LX1869	LX2R11C15Z4S	Luton & District	407	H407ERO	Leyland	DP29F	10/90
		The Shires	407	H407ERO	Leyland	DP29F	04/95
LX1870	LX2R11C15Z4S	Luton & District	408	H408ERO	Leyland	DP29F	10/90
		The Shires	408	H408ERO	Leyland	DP45F	04/95
		The Shires	408	H408ERO	Leyland	DP29F	06/95
LX1871	LX2R11C15Z4S	Luton & District	409	H409ERO	Leyland	DP29F	10/90
		The Shires	409	F409ERO	Leyland	DP29F	04/95
LX1872	LX2R11C15Z4S	Luton & District	410	H410ERO	Leyland	DP29F	10/90
		The Shires	410	F410ERO	Leyland	DP29F	04/95
		The Shires	410	F410ERO	Leyland	DP45F	06/95
LX1873	LX2R11C15Z4R	City Line	1634	H634YHT	Leyland	B49F	08/90
LX1874	LX2R11C15Z4R	City Line	1636	H636YHT	Leyland	B49F	08/90
LX1875	LX2R11C15Z4R	City Line	1637	H637YHT	Leyland	B49F	08/90
LX1876	LX2R11C15Z4R	City Line	1638	H638YHT	Leyland	B49F	08/90
LX1877	LX2R11C15Z4R	City Line	1639	H639YHT	Leyland	B49F	08/90
LX1878	LX2R11C15Z4R	City Line	1640	H640YHT	Leyland	B49F	08/90
LX1879	LX2R11C15Z4R	City Line	1641	H641YHT	Leyland	B49F	08/90
LX1880	LX2R11C15Z4R	City Line	1642	H642YHT	Leyland	B49F	08/90
LX1881	LX2R11C15Z4R	City Line	1643	H643YHT	Leyland	B49F	08/90
LX1882	LX2R11C15Z4R	City Line	1644	H644YHT	Leyland	B49F	08/90
LX1883	LX2R11C15Z4R	City Line	1645	H645YHT	Leyland	B49F	08/90
LX1884	LX2R11C15Z4R	City Line	1646	H646YHT	Leyland	B49F	08/90
LX1885	LX2R11C15Z4R	City Line	1647	H647YHT	Leyland	B49F	08/90
LX1886	LX2R11C15Z4R	City Line	1648	H648YHT	Leyland	B49F	08/90
LX1887	LX2R11C15Z4R	City Line	1649	H649YHT	Leyland	B49F	08/90

Badgerline's own fleet operates seven Lynx from its depot at Wells. Shown here while heading for Yeovil is 3615, H615YTC. *Malc McDonald*

LX1888	LX2R11C15Z4R	City Line	1650	H650YHT	Leyland	B49F	08/90
LX1889	LX2R11C15Z4R	City Line	1651	H651YHT	Leyland	B49F	08/90
LX1890	LX2R11C15Z4R	West Midlands Travel	1316	G316EOG	Leyland	B49F	08/90
LX1891	LX2R11C15Z4R	City Line	1652	H652YHT	Leyland	B49F	08/90
LX1892	LX2R11C15Z4R	City Line	1653	H653YHT	Leyland	B49F	08/90
LX1893	LX2R11C15Z4R	City Line	1654	H654YHT	Leyland	B49F	08/90
LX1894	LX2R11C15Z4R	City Line	1655	H655YHT	Leyland	B49F	08/90
LX1895	LX2R11C15Z4R	City Line	1656	H656YHT	Leyland	B49F	08/90
LX1896	LX2R11C15Z4R	City Line	1657	H657YHT	Leyland	B49F	08/90
LX1897	LX2R11C15Z4R	City Line	1658	H658YHT	Leyland	B49F	08/90
LX1898	LX2R11C15Z4R	City Line	1659	H659YHT	Leyland	B49F	08/90
LX1899	LX2R11C15Z4R	City Line	1660	H660YHT	Leyland	B49F	08/90
LX1900	LX2R11C15Z4R	City Line	1661	H661YHT	Leyland	B49F	08/90
LX1901	LX2R11C15Z4R	City Line	1662	H662YHT	Leyland	B49F	08/90
LX1902	LX2R11C15Z4R	Badgerline	3610	H610YTC	Leyland	B49F	08/90
LX1903	LX2R11C15Z4R	Badgerline	3611	H611YTC	Leyland	B49F	08/90
LX1904	LX2R11C15Z4R	Badgerline	3612	H612YTC	Leyland	B49F	08/90
LX1905	LX2R11C15Z4R	Badgerline	3613	H613YTC	Leyland	B49F	08/90
LX1906	LX2R11C15Z4R	Badgerline	3614	H614YTC	Leyland	B49F	08/90
LX1907	LX2R11C15Z4R	Badgerline	3615	H615YTC	Leyland	B49F	08/90
LX1908	LX2R11C15Z4R	Badgerline	3616	H616YTC	Leyland	B49F	08/90
LX1909	LX2R11C15Z4S	Busways	126	H126ACU	Leyland	DP47F	08/90
LX1910	LX2R11C15Z4S	Busways	127	H127ACU	Leyland	DP47F	08/90

LYNX II PRODUCTION

LX2001	LX2R11C15Z4R	VL Bus				Leyland	B51F	06/90
LX2002	LX2R11C15Z4R	VL Bus			H48NDU	Leyland	B51F	07/90
		Hedingham		L196	H48NDU	Leyland	B51F	02/92
LX2003	LX2R11C15Z4S	VL Demonstrator			H733HWK	Leyland	B51F	11/90
		McGill's Bus Service			H733HWK	Leyland	B51F	07/91
LX2004	LX2R11V18Z4R	The Wright Company			H256YLG	Leyland	B49F	12/90
		Aintree Coachline			H256YLG	Leyland	B49F	02/94
LX2005	LX2R11C15Z4R	Halton		50	H543FWN	Leyland	B51F	10/90
LX2006	LX2R11C15Z4R	Halton		35	H35HBG	Leyland	B51F	05/91
LX2007	LX2R11C15Z4R	Halton		51	H544FWN	Leyland	B51F	10/90
LX2008	LX2R11C15Z4R	Colchester		30	H130LPU	Leyland	B49F	12/90
		Crosville Wales		SLC30	H130LPU	Leyland	B49F	08/94
LX2009	LX2R11C15Z4S	West Riding		333	H338TYG	Leyland	B49F	11/90
LX2010	LX2R11C15Z4S	West Riding		334	H334TYG	Leyland	B49F	11/90
LX2011	LX2R11C15Z4S	West Riding		335	H335TYG	Leyland	B49F	11/90
LX2012	LX2R11C15Z4S	West Riding		336	H336TYG	Leyland	B49F	11/90
LX2013	LX2R11C15Z4S	West Riding		337	H337TYG	Leyland	B49F	11/90
LX2014	LX2R11C15Z4S	West Riding		338	H338UWT	Leyland	B49F	12/90
LX2015	LX2R11C15Z4S	West Riding		339	H339UWT	Leyland	B49F	12/90
LX2016	LX2R11C15Z4S	West Riding		340	H343UWT	Leyland	B49F	12/90
LX2017	LX2R11C15Z4S	West Riding		341	H341UWT	Leyland	B49F	12/90
LX2018	LX2R11C15Z4S	West Riding		342	H342UWT	Leyland	B49F	12/90
LX2019	LX2R11C15Z4S	VL demonstrator			H49NDU	Leyland	B49F	06/91
		Cardiff Bus		260	H49NDU	Leyland	B49F	10/91
LX2020	LX2R11C15Z4S	VL demonstrator			H51NDU	Leyland	B49F	07/91
		Whitelaw's, Stonehouse			H51NDU	Leyland	B49F	07/91
		Liverbus			H51NDU	Leyland	B49F	09/93
		Redby Travel, Sunderland			H51NDU	Leyland	B49F	12/93
LX2021	LX2R11C15Z4S	Whitelaw's, Stonehouse			J41GGB	Leyland	B49F	08/91
		Dursley, Silvertown			J41GGB	Leyland	B49F	08/93
		Redby Travel, Sunderland			J41GGB	Leyland	B49F	09/94
LX2022	LX2R11C15Z4S	Whitelaw's, Stonehouse			J42GGB	Leyland	B49F	08/91
		Dursley, Silvertown			J42GGB	Leyland	B49F	08/93
		Redby Travel, Sunderland			J42GGB	Leyland	B51F	09/94
LX2023	LX2R11C15Z4S	Cardiff Bus		261	J261UDW	Leyland	B49F	11/91
LX2024	LX2R11C15Z4S	Whitelaw's, Stonehouse			J43GGB	Leyland	B49F	08/91
		Redby Travel, Sunderland			J42GGB	Leyland	B51F	12/93
LX2025	LX2R11C15Z4S	Cardiff Bus		262	J262UDW	Leyland	B49F	11/91
LX2026	LX2R11C15Z4S	Cardiff Bus		263	J263UDW	Leyland	B49F	11/91
LX2027	LX2R11V18Z4R	City of Nottingham		750	H47NDU	Leyland	B50F	06/91
LX2028	LX2R11C15Z4S	West Riding		343	H343UWX	Leyland	B49F	10/91
LX2029	LX2R11C15Z4S	West Riding		344	H344UWX	Leyland	B49F	10/91
LX2030	LX2R11C15Z4S	West Riding		345	H345UWX	Leyland	B49F	10/91
LX2031	LX2R11C15Z4S	West Riding		346	H346UWX	Leyland	B49F	10/91
LX2032	LX2R11C15Z4S	West Riding		347	H347UWX	Leyland	B49F	10/91
LX2033	LX2R11C15Z4S	Cardiff Bus		264	J264UDW	Leyland	B49F	11/91
LX2034	LX2R11C15Z4S	Cardiff Bus		265	J265UDW	Leyland	B49F	11/91
LX2035	LX2R11C15Z4S	Cardiff Bus		266	J266UDW	Leyland	B49F	11/91
LX2036	LX2R11C15Z4S	Cardiff Bus		267	J267UDW	Leyland	B49F	11/91
LX2037	LX2R11C15Z4S	Volvo Bus				Leyland	B F	11/91
		Body development vehicle dismantled						*01/93*
LX2038	LX2R11C15Z4S	Cardiff Bus		268	J268UDW	Leyland	B49F	11/91
LX2039	LX2R11C15Z4S	Cardiff Bus		269	J269UDW	Leyland	B49F	11/91
LX2040	LX2R11C15Z4R	John Fishwick & Sons		4	H64CCK	Leyland	B47F	04/91

The Lynx II introduced many changes from the earlier version, the most noticeable being the protruding front introduced to cater for the Volvo enhancements. Shown here are, above, H256YLG of The Wright Company of Wrexham, photographed leaving Chester for Connah's Quay and below, K27EWC in Colchester Borough Transport livery. Note that the Colchester vehicle included roof driving lights introduced shortly after the mark II went into production. *Richard Eversden/Keith Grimes*

Lynx IIs for Scotland are represented by two vehicles in independents' liveries. McGill's Bus Services is a long established family bus company operating from Barrhead and its H733HWK is seen in that town when new. The vehicle features a Cummins L10 engine, ZF 4HP500 gearbox and a stepped floor. Whitelaw's Lynx J45GGB had been placed for sale when photographed in Liverpool in September 1993 working with Liverbus. Whitelaw's took three original versions of the Lynx and five of the Lynx IIs, though these have now been released for service elsewhere. *Volvo Bus/Keith Grimes*

LX2041	LX2R11C15Z4R	John Fishwick & Sons	5	H65CCK	Leyland	B51F	04/91
LX2042	LX2R11C15Z4R	Halton	36	J249KWM	Leyland	B51F	08/91
LX2043	LX2R11C15Z4R	John Fishwick & Sons	7	J7JFS	Leyland	B51F	01/92
LX2044	LX2R11C15Z4R	John Fishwick & Sons	14	J14JFS	Leyland	B51F	03/92
LX2045	LX2R11C15Z4S	Cardiff Bus	270	J270UDW	Leyland	B49F	11/91
LX2046	LX2R11C15Z4S	Cardiff Bus	271	J271UDW	Leyland	B49F	11/91
LX2047	LX2R11C15Z4S	Whitelaw's, Stonehouse		J45GGB	Leyland	B51F	03/92
		Liverbus		J45GGB	Leyland	B51F	09/93
		Tellings-Golden Miller		J45GGB	Leyland	B51F	12/93
		Redby Travel, Sunderland		J45GGB	Leyland	B51F	09/94
LX2048	LX2R11C15Z4R	Colchester	27	K27EWC	Leyland	B51F	09/92
		Crosville Wales	SLC27	K27EWC	Leyland	B51F	08/94
LX2049	LX2R11C15Z4S	Volvo Bus		J295TWK	Leyland	B51F	02/92
		Hedingham		J295TWK	Leyland	B51F	07/92
LX2050	LX2R11C15Z4S	Volvo Bus		J564URW	Leyland	B51F	03/92
		Tees	5050	J564URW	Leyland	B51F	03/92
		Felix, Stanley		J564URW	Leyland	B51F	11/92
LX2051	LX2R11C15Z4S	Halton	59	J922MKC	Leyland	B51F	05/92
LX2052	LX2R11C15Z4S	Halton	58	J921MKC	Leyland	B51F	05/92
LX2053	LX2R11C15Z4S	Lothian	177	H177OSG	Leyland	B43D	03/91
LX2054	LX2R11C15Z4S	Lothian	178	H178OSG	Leyland	B43D	03/91
LX2055	LX2R11C15Z4S	Lothian	179	H179OSG	Leyland	B43D	03/91
LX2056	LX2R11C15Z4S	Lothian	180	H180OSG	Leyland	B43D	03/91
LX2057	LX2R11C15Z4S	Lothian	181	H181OSG	Leyland	B43D	03/91
LX2058	LX2R11C15Z4S	Lothian	182	H182OSG	Leyland	B43D	03/91
LX2059	LX2R11C15Z4S	Lothian	183	H183OSG	Leyland	B43D	03/91
LX2060	LX2R11C15Z4S	Lothian	184	H184OSG	Leyland	B43D	03/91
LX2061	LX2R11C15Z4S	Lothian	185	H185OSG	Leyland	B43D	03/91
LX2062	LX2R11C15Z4S	Lothian	186	H186OSG	Leyland	B43D	03/91
LX2063	LX2R11C15Z4S	Lothian	187	H187OSG	Leyland	B43D	03/91
LX2064	LX2R11C15Z4S	Lothian	188	H188OSG	Leyland	B43D	03/91
LX2065	LX2R11C15Z4R	Halton	46	J628LHF	Leyland	B51F	01/92
LX2066	LX2R11C15Z4R	Halton	38	J251KWM	Leyland	B51F	11/91
LX2067	LX2R11C15Z4R	Halton	53	K852MTJ	Leyland	B51F	08/92
LX2068	LX2R11C15Z4R	Halton	48	J629LHF	Leyland	B51F	02/92
LX2069	LX2R11C15Z4R	Halton	61	J924MKC	Leyland	B51F	06/92
LX2070	LX2R11C15Z4R	Halton	62	J926MKC	Leyland	B51F	06/92
LX2071	LX2R11C15Z4R	Halton	60	J923MKC	Leyland	B51F	06/92
LX2072	LX2R11C15Z4R	Halton	64	J928MKC	Leyland	B51F	07/92
LX2073	LX2R11C15Z4R	Halton	63	J927MKC	Leyland	B51F	07/92
LX2074	LX2R11C15Z4S	Yorkshire Buses	352	H755WWW	Leyland	B49F	06/91
LX2075	LX2R11C15Z4S	Yorkshire Buses	353	H756WWW	Leyland	B49F	06/91
LX2076	LX2R11C15Z4S	Yorkshire Buses	354	H757WWW	Leyland	B49F	06/91
LX2077	LX2R11C15Z4S	Yorkshire Buses	355	H355WWX	Leyland	B49F	07/91
LX2078	LX2R11C15Z4S	Yorkshire Buses	356	H356WWX	Leyland	B49F	07/91
LX2079	LX2R11C15Z4S	Yorkshire Buses	357	H357WWX	Leyland	B49F	07/91
LX2080	LX2R11C15Z4S	West Riding	358	H358WWY	Leyland	B49F	07/91
LX2081	LX2R11C15Z4S	West Riding	359	H359WWY	Leyland	B49F	07/91
LX2082	LX2R11C15Z4S	Selby & District	360	H460WWY	Leyland	B49F	07/91
		West Riding	360	H460WWY	Leyland	B49F	03/93
LX2083	LX2R11C15Z4R	Selby & District	361	H393WWY	Leyland	B49F	07/91
		West Riding	361	H393WWY	Leyland	B49F	03/93

Opposite: **Lynx IIs from the capitals are seen opposite. From the capital of Scotland one of the only batch of dual-doored Lynx remaining in service is Lothian 180, H180OSG. These vehicles regularly featured in the Lynx promotion material from 1991 to the end of marketing. Shown in the Welsh capital's white and orange scheme is Cardiff Bus 260, H49NDU after a period of operation as a demonstrator for the model.** *Tony Wilson/David Cole*

LX2084	LX2R11C15Z4S	West Riding	362	J362YWX	Leyland	B49F	08/91
LX2085	LX2R11C15Z4S	West Riding	363	J363YWX	Leyland	B49F	08/91
LX2086	LX2R11C15Z4S	West Riding	364	J364YWX	Leyland	B49F	08/91
LX2087	LX2R11C15Z4S	West Riding	365	J365YWX	Leyland	B49F	08/91
LX2088	LX2R11C15Z4S	United	5006	H31PAJ	Leyland	B49F	06/91
LX2089	LX2R11C15Z4S	Tees	5007	H32PAJ	Leyland	B49F	06/91
LX2090	LX2R11C15Z4S	Tees	5008	H34PAJ	Leyland	B49F	06/91
LX2091	LX2R11C15Z4S	Tees	5009	H253PAJ	Leyland	B49F	07/91
LX2092	LX2R11C15Z4S	Tees	5010	H254PAJ	Leyland	B49F	07/91
LX2093	LX2R11C15Z4S	Yorkshire Buses	366	J366YWX	Leyland	B49F	09/91
LX2094	LX2R11C15Z4S	Yorkshire Buses	367	J367YWX	Leyland	B49F	09/91
LX2095	LX2R11C15Z4S	Yorkshire Buses	368	J368YWX	Leyland	B49F	09/91
LX2096	LX2R11C15Z4S	Yorkshire Buses	369	J369YWX	Leyland	B49F	10/91
LX2097	LX2R11C15Z4S	Yorkshire Buses	370	J370YWX	Leyland	B49F	10/91
LX2098	LX2R11C15Z4S	Yorkshire Buses	371	J371YWX	Leyland	B49F	10/91
LX2099	LX2R11C15Z4S	West Riding	372	J372AWT	Leyland	B49F	10/91
LX2100	LX2R11C15Z4S	West Riding	373	J373AWT	Leyland	B49F	10/91
LX2101	LX2R11C15Z4S	West Riding	374	J374AWT	Leyland	B49F	10/91
LX2102	LX2R11C15Z4S	West Riding	375	J375AWT	Leyland	B49F	11/91
LX2103	LX2R11C15Z4S	West Riding	376	J376AWT	Leyland	B49F	11/91
LX2104	LX2R11C15Z4S	West Riding	377	J377AWT	Leyland	B49F	11/91
LX2105	LX2R11C15Z4S	United	5011	J651UHN	Leyland	B49F	10/91
LX2106	LX2R11C15Z4S	United	5012	J652UHN	Leyland	B49F	10/91
LX2107	LX2R11C15Z4S	United	5013	J653UHN	Leyland	B49F	10/91
LX2108	LX2R11C15Z4S	Tees	5014	J654UHN	Leyland	B49F	10/91
LX2109	LX2R11C15Z4S	Tees	5015	J655UHN	Leyland	B49F	10/91
LX2110	LX2R11V18Z4S	West Riding	378	J371AWT	Leyland	B49F	12/91
LX2111	LX2R11C15Z4S	Tees	5016	J656UHN	Leyland	B49F	12/91
LX2112	LX2R11C15Z4S	United	5017	J657UHN	Leyland	B49F	12/91
LX2113	LX2R11C15Z4S	United	5018	J658UHN	Leyland	B49F	12/91
LX2114	LX2R11C15Z4S	West Riding	379	J379BWU	Leyland	B49F	12/91
LX2115	LX2R11C15Z4S	West Riding	380	J380BWU	Leyland	B49F	12/91
LX2116	LX2R11C15Z4S	West Riding	381	J381BWU	Leyland	B49F	01/92
LX2117	LX2R11C15Z4S	West Riding	382	J382BWU	Leyland	B49F	01/92
LX2118	LX2R11V18Z4S	Volvo Bus		J901UKV	Leyland	B51F	03/92
		Cleveland Transit	21	J901UKV	Leyland	B51F	08/92
LX2119	LX2R11V18Z4S	City of Nottingham	759	J759DAU	Leyland	B50F	01/92
LX2120	LX2R11V18Z4S	City of Nottingham	760	J760DAU	Leyland	B50F	01/92
LX2121	LX2R11V18Z4S	Westbus, Ashford		J724KBC	Leyland	B51F	05/92
		Hedingham	L206	J724KBC	Leyland	B51F	02/94
LX2122	LX2R11V18Z4S	Cleveland Transit	22	K622YVN	Leyland	B51F	08/92
LX2123	LX2R11V18Z4S	Volvo Bus		J916WVC	Leyland	B51F	06/92
		Brewers	508	J916WVC	Leyland	B51F	08/92
		Brewers	508	J916WVC	Leyland	B47F	04/93
LX2124	LX2R11V18Z4S	Cleveland Transit	23	K623YVN	Leyland	B51F	08/92
LX2125	LX2R11V18Z4S	Cleveland Transit	24	K624YVN	Leyland	B51F	08/92
LX2126	LX2R11V18Z4S	Cleveland Transit	25	K625YVN	Leyland	B51F	08/92
LX2127	LX2R11V18Z4S	Cleveland Transit	26	K626YVN	Leyland	B51F	08/92
LX2128	LX2R11V18Z4S	Cleveland Transit	27	K627YVN	Leyland	B51F	08/92
LX2129	LX2R11V18Z4S	Cleveland Transit	28	K628YVN	Leyland	B51F	08/92
LX2130	LX2R11V18Z4S	Cleveland Transit	29	K629YVN	Leyland	B51F	08/92
LX2131	LX2R11V18Z4S	Cleveland Transit	30	K630YVN	Leyland	B51F	08/92
LX2132	LX2R11V18Z4S	Volvo Bus		J375WWK	Leyland	B51F	07/92
		Brewers	509	J375WWK	Leyland	B51F	08/92
		Brewers	509	J375WWK	Leyland	B47F	04/93

Brewers took one of the last Lynx as its 509, J375WWK. The vehicle is seen in Bridgend in February 1993 lettered for the Cymmer to Bridgend service, though operating the 232 to Glyncorrwg at the time. LX2132 was used for demonstration work in 1992 and was joined at Brewers by three other Lynx IIs. Since delivery, these have been modified to B47F configuration with additional luggage space and have also lost the dedicated service lettering. *Richard Eversden*

J564URW was another of the Lynx to spend some time on demonstration work before being based with North East Buses' Tees operation between March and September 1992. The bus is now operated by Felix on its Derby to Ilkeston service. *Malc McDonald*

The last Lynx to enter service were a pair for Alder Valley. The service to which they were allocated was sold to Stagecoach South a month later, though they returned to the Q-Drive Group in January 1993. K801CAN has since been painted into The Bee Line's yellow livery. *Malc McDonald*

LX2133	LX2R11V18Z4S	Volvo Training School			Leyland	B51F	07/92
		Metrobus, Orpington		K101JMV	Leyland	B51F	08/92
LX2134	LX2R11V18Z4S	Alder Valley	801	K801CAN	Leyland	B51F	09/92
		Stagecoach South	701	K801CAN	Leyland	B51F	10/92
		The Bee Line	801	K801CAN	Leyland	B51F	01/93
LX2135	LX2R11V18Z4S	Alder Valley	802	K802CAN	Leyland	B51F	09/92
		Stagecoach South	702	K802CAN	Leyland	B51F	10/92
		The Bee Line	802	K802CAN	Leyland	B51F	01/93
LX2136	LX2R11C15Z4R	Brewers	510	K10BMB	Leyland	B51F	09/92
		Brewers	510	K10BMB	Leyland	B47F	04/93
LX2137	LX2R11C15Z4R	Brewers	511	K11BMB	Leyland	B51F	09/92
		Brewers	511	K11BMB	Leyland	B47F	04/93
LX2138	LX2R11C15Z4R	Brewers	512	K12BMB	Leyland	B51F	09/92
		Brewers	512	K12BMB	Leyland	B47F	04/93
LX2139	LX2R11C15Z4R	Halton	47	J630LHF	Leyland	B49F	04/92
LX2140	LX2R11C15Z4R	Halton	49	J929MKC	Leyland	B49F	04/92
LX2141	LX2R11C15Z4R	Halton	52	J925MKC	Leyland	B49F	06/92
LX2142	LX2R11C15Z4R	Halton	57	K853MTJ	Leyland	B49F	08/92

Opposite top: **Metrobus of Orpington operates several Lynx including three from Merthyr Tydfil and a pair from Miller of Foxton. LX2133, the only Lynx II example, arrived in Orpington after a period with the Volvo training school.** *Tony Wilson*
Opposite, bottom: **The final Lynx off the assembly line followed a tradition at the plant by going to Halton. Halton had the honour of many notable firsts and lasts from the Lillyhall plant including the first single-doored National, the last National and the first Lynx II into service. Seen here is 57, K853MTJ.**

Index

One of the most significant orders for the Lynx was the 250 for West Midlands Travel. Seen here prior to being formally handed over to the operator is, LX1467 the first of the main production run. In 1995, West Midlands Travel ordered 300 of the Lynx successor, the Volvo B10B and these are expected to enter service commencing early 1996.

F313AWW	LX1412	F428MJN	LX1206	F619RTC	LX1462	G37VME	LX1507
F314AWW	LX1225	F429MJN	LX1207	F620RTC	LX1463	G38VME	LX1508
F358JVS	LX1116	F520AEM	LX1333	F621RTC	LX1464	G38YHJ	LX1622
F359JVS	LX1115	F521AEM	LX1384	F622RTC	LX1465	G39VME	LX1509
F361YTJ	LX1169	F538LUF	LX1234	F623RTC	LX1466	G39YHJ	LX1623
F362YTJ	LX1170	F544LUF	LX1197	F624RTC	LX1479	G40VME	LX1524
F363YTJ	LX1171	F545LUF	LX1261	F625RTC	LX1480	G40YHJ	LX1624
F364YTJ	LX1196	F546LUF	LX1260	F626RTC	LX1481	G41VME	LX1525
F400PUR	LX1313	F556NJM	LX1218	F627RTC	LX1490	G42VME	LX1538
F401PUR	LX1328	F557NJM	LX1219	F628RTC	LX1491	G43VME	LX1539
F402LTW	LX1160	F558NJM	LX1220	F629RTC	LX1492	G44VME	LX1540
F402PUR	LX1329	F559NJM	LX1232	F630RTC	LX1493	G45VME	LX1541
F403LTW	LX1162	F560NJM	LX1233	F631RTC	LX1494	G49CVC	LX1317
F403PUR	LX1330	F600RTC	LX1390	F632RTC	LX1495	G73UYV	LX1478
F404LTW	LX1163	F601RTC	LX1444	F660PWK	LX1155	G74UYV	LX1519
F404PUR	LX1332	F601UVN	LX1334	F661PWK	LX1154	G75UYV	LX1520
F405LTW	LX1164	F602RTC	LX1445	F687YWM	LX1267	G76UYV	LX1521
F406LTW	LX1172	F602UVN	LX1335	F722LRG	LX1243	G77UYV	LX1522
F407LTW	LX1173	F603RTC	LX1446	F723LRG	LX1245	G78UYV	LX1523
F408LTW	LX1174	F603UVN	LX1348	F724LRG	LX1246	G79EOG	LX1482
F409LTW	LX1175	F604RTC	LX1447	F725LRG	LX1247	G80EOG	LX1483
F410MNO	LX1176	F604UVN	LX1349	F726LRG	LX1248	G81EOG	LX1484
F411MNO	LX1177	F605RTC	LX1448	F727LRG	LX1244	G82EOG	LX1485
F412MNO	LX1178	F605UVN	LX1357	F728LRG	LX1135	G83EOG	LX1486
F413MNO	LX1179	F606RTC	LX1449	F729LRG	LX1251	G84EOG	LX1487
F414MNO	LX1180	F607RTC	LX1450	F730LRG	LX1250	G85EOG	LX1488
F415MNO	LX1181	F608RTC	LX1451	F731LRG	LX1137	G86EOG	LX1489
F416MNO	LX1186	F608WBV	LX1215	F732LRG	LX1249	G87EOG	LX1496
F417MNO	LX1187	F609RTC	LX1452	F733LRG	LX1268	G88EOG	LX1497
F418MNO	LX1190	F610RTC	LX1453	F740HRC	LX1297	G89EOG	LX1498
F419MNO	LX1191	F610UVN	LX1358	F741HRC	LX1298	G90EOG	LX1499
F420MJN	LX1198	F611RTC	LX1454	F742HRC	LX1299	G91EOG	LX1500
F421MJN	LX1199	F612RTC	LX1455	F743HRC	LX1300	G92EOG	LX1501
F422MJN	LX1200	F613RTC	LX1456	F744HRC	LX1301	G93EOG	LX1502
F423MJN	LX1201	F614RTC	LX1457	F883SMU	LX1152	G94EOG	LX1503
F424MJN	LX1202	F615RTC	LX1458	F895BKF	LX1440	G95EOG	LX1511
F425MJN	LX1203	F616RTC	LX1459	G34VME	LX1504	G96EOG	LX1512
F426MJN	LX1204	F617RTC	LX1460	G35VME	LX1505	G97EOG	LX1513
F427MJN	LX1205	F618RTC	LX1461	G36VME	LX1506	G98EOG	LX1514

G99EOG	LX1515	G131HNP	LX1748	G186EOG	LX1653	G248EOG	LX1802
G100EOG	LX1516	G132EOG	LX1565	G187EOG	LX1654	G249EOG	LX1803
G101EOG	LX1517	G132HNP	LX1749	G188EOG	LX1655	G249HUH	LX1402
G101HNP	LX1679	G133EOG	LX1566	G189EOG	LX1656	G250EOG	LX1804
G102EOG	LX1518	G133HNP	LX1750	G190EOG	LX1657	G250HUH	LX1403
G102HNP	LX1696	G134EOG	LX1567	G191EOG	LX1658	G251EOG	LX1805
G103EOG	LX1526	G134HNP	LX1762	G192EOG	LX1659	G251HUH	LX1423
G103HNP	LX1697	G135EOG	LX1568	G193EOG	LX1660	G252EOG	LX1806
G104EOG	LX1527	G135HNP	LX1763	G194EOG	LX1661	G252HUH	LX1434
G104HNP	LX1698	G136EOG	LX1569	G195EOG	LX1662	G253EOG	LX1807
G104WRV	LX1378	G136HNP	LX1764	G196EOG	LX1663	G253HUH	LX1435
G105EOG	LX1528	G137EOG	LX1570	G197EOG	LX1664	G254EOG	LX1808
G105HNP	LX1699	G137HNP	LX1765	G198EOG	LX1665	G254HUH	LX1437
G105WRV	LX1379	G138EOG	LX1571	G199EOG	LX1666	G255EOG	LX1689
G106EOG	LX1529	G138HNP	LX1766	G200EOG	LX1667	G255HUH	LX1438
G106HNP	LX1704	G139EOG	LX1572	G201EOG	LX1668	G256EOG	LX1731
G106WRV	LX1385	G139HNP	LX1767	G201URO	LX1574	G256HUH	LX1439
G107EOG	LX1530	G140EOG	LX1573	G202EOG	LX1669	G257EOG	LX1732
G107HNP	LX1705	G140HNP	LX1768	G202URO	LX1575	G257HUH	LX1441
G107WRV	LX1386	G141EOG	LX1581	G203EOG	LX1670	G258EOG	LX1733
G108EOG	LX1531	G141HNP	LX1771	G203URO	LX1576	G258HUH	LX1442
G108HNP	LX1706	G142EOG	LX1582	G204EOG	LX1671	G259EOG	LX1734
G108OUG	LX1629	G142HNP	LX1772	G204URO	LX1577	G259HUH	LX1443
G108WRV	LX1387	G143EOG	LX1583	G205EOG	LX1672	G260EOG	LX1742
G109EOG	LX1532	G143HNP	LX1773	G205URO	LX1578	G261EOG	LX1743
G109HNP	LX1707	G144EOG	LX1584	G206EOG	LX1673	G261LUG	LX1306
G109OUG	LX1604	G144HNP	LX1774	G206URO	LX1579	G262EOG	LX1744
G109WRV	LX1388	G145EOG	LX1585	G207EOG	LX1680	G263EOG	LX1751
G110EOG	LX1533	G145HNP	LX1775	G207URO	LX1580	G264EOG	LX1752
G110HNP	LX1708	G146EOG	LX1586	G208EOG	LX1681	G265EOG	LX1753
G110OUG	LX1631	G146HNP	LX1776	G209EOG	LX1682	G266EOG	LX1754
G110WRV	LX1389	G147EOG	LX1587	G210EOG	LX1683	G267EOG	LX1755
G111EOG	LX1534	G147HNP	LX1778	G211EOG	LX1684	G268EOG	LX1756
G111HNP	LX1709	G148CHP	LX1393	G212EOG	LX1685	G269EOG	LX1757
G111XOW	LX1391	G148EOG	LX1588	G213EOG	LX1686	G270EOG	LX1809
G112EOG	LX1535	G148HNP	LX1779	G214EOG	LX1700	G271EOG	LX1810
G112HNP	LX1710	G149CHP	LX1392	G214KRN	LX1606	G272EOG	LX1811
G112XOW	LX1308	G149EOG	LX1589	G215EOG	LX1701	G273EOG	LX1812
G113EOG	LX1536	G149HNP	LX1780	G215KRN	LX1607	G274EOG	LX1813
G113HNP	LX1711	G150EOG	LX1590	G216EOG	LX1702	G275EOG	LX1814
G113XOW	LX1309	G150HNP	LX1781	G216KRN	LX1608	G276EOG	LX1815
G114EOG	LX1542	G151EOG	LX1591	G217EOG	LX1703	G277EOG	LX1816
G114HNP	LX1719	G152EOG	LX1592	G217KRN	LX1609	G278EOG	LX1817
G115EOG	LX1543	G153EOG	LX1593	G218EOG	LX1712	G279EOG	LX1818
G115HNP	LX1720	G154EOG	LX1594	G218KRN	LX1610	G280EOG	LX1819
G116EOG	LX1544	G155EOG	LX1595	G219EOG	LX1713	G281EOG	LX1820
G116HNP	LX1721	G156EOG	LX1596	G220EOG	LX1714	G282EOG	LX1821
G117EOG	LX1545	G157EOG	LX1597	G221DKA	LX1510	G283EOG	LX1822
G117HNP	LX1722	G158EOG	LX1611	G221EOG	LX1715	G284EOG	LX1823
G118EOG	LX1546	G159EOG	LX1612	G222DKA	LX1537	G285EOG	LX1824
G118HNP	LX1723	G160EOG	LX1613	G222EOG	LX1716	G286EOG	LX1825
G119EOG	LX1547	G161EOG	LX1614	G223EOG	LX1717	G287EOG	LX1826
G119HNP	LX1724	G162EOG	LX1615	G224EOG	LX1718	G288EOG	LX1827
G120EOG	LX1548	G163EOG	LX1616	G225EOG	LX1726	G289EOG	LX1828
G120HNP	LX1725	G164EOG	LX1617	G226EOG	LX1727	G290EOG	LX1829
G121EOG	LX1549	G165EOG	LX1618	G227EOG	LX1728	G291EOG	LX1830
G121HNP	LX1735	G166EOG	LX1619	G228EOG	LX1729	G292EOG	LX1831
G122EOG	LX1550	G167EOG	LX1620	G229EOG	LX1730	G293EOG	LX1832
G122HNP	LX1736	G168EOG	LX1621	G230EOG	LX1760	G293KWY	LX1314
G123EOG	LX1551	G169EOG	LX1625	G231EOG	LX1761	G294EOG	LX1833
G123HNP	LX1737	G170EOG	LX1626	G232EOG	LX1782	G294KWY	LX1315
G124EOG	LX1552	G171EOG	LX1627	G233EOG	LX1783	G295EOG	LX1834
G124HNP	LX1738	G172EOG	LX1628	G234EOG	LX1784	G295KWY	LX1320
G125EOG	LX1553	G173EOG	LX1635	G235EOG	LX1785	G296EOG	LX1835
G125HNP	LX1739	G174EOG	LX1636	G236EOG	LX1786	G296KWY	LX1321
G126EOG	LX1554	G175EOG	LX1637	G237EOG	LX1787	G297EOG	LX1836
G126HNP	LX1740	G176EOG	LX1638	G238EOG	LX1788	G297KWY	LX1322
G127EOG	LX1555	G177EOG	LX1639	G239EOG	LX1789	G298EOG	LX1837
G127HNP	LX1741	G178EOG	LX1640	G240EOG	LX1790	G298KWY	LX1323
G128EOG	LX1556	G179EOG	LX1641	G241EOG	LX1791	G299EOG	LX1838
G128HNP	LX1745	G180EOG	LX1642	G242EOG	LX1792	G299KWY	LX1324
G129EOG	LX1557	G181EOG	LX1643	G243EOG	LX1793	G300EOG	LX1839
G129HNP	LX1746	G182EOG	LX1644	G244EOG	LX1794	G300KWY	LX1325
G130EOG	LX1558	G183EOG	LX1650	G245EOG	LX1795	G301EOG	LX1840
G130HNP	LX1747	G184EOG	LX1651	G246EOG	LX1800	G302EOG	LX1841
G131EOG	LX1564	G185EOG	LX1652	G247EOG	LX1801	G303EOG	LX1843

G304EOG	LX1844	H32PAJ	LX2089	H636YHT	LX1874	J363YWX	LX2085
G305EOG	LX1845	H34HBG	LX1274	H637YHT	LX1875	J364YWX	LX2086
G306EOG	LX1846	H34PAJ	LX2090	H638YHT	LX1876	J365YWX	LX2087
G307EOG	LX1847	H35HBG	LX2006	H639YHT	LX1877	J366YWX	LX2093
G308EOG	LX1849	H47NDU	LX2027	H640YHT	LX1878	J367YWX	LX2094
G309EOG	LX1850	H48NDU	LX2002	H641YHT	LX1879	J368YWX	LX2095
G310EOG	LX1851	H49NDU	LX2019	H642YHT	LX1880	J369YWX	LX2096
G311DPA	LX1605	H51NDU	LX2020	H643YHT	LX1881	J370YWX	LX2097
G311EOG	LX1852	H64CCK	LX2040	H644YHT	LX1882	J371AWT	LX2110
G312DPA	LX1674	H65CCK	LX2041	H645YHT	LX1883	J371YWX	LX2098
G312EOG	LX1853	H126ACU	LX1909	H646YHT	LX1884	J372AWT	LX2099
G313DPA	LX1675	H127ACU	LX1910	H647YHT	LX1885	J373AWT	LX2100
G313EOG	LX1854	H130LPU	LX2008	H648YHT	LX1886	J374AWT	LX2101
G314DPA	LX1676	H177OSG	LX2053	H649YHT	LX1887	J375AWT	LX2102
G314EOG	LX1855	H178OSG	LX2054	H650YHT	LX1888	J375WWK	LX2132
G315DPA	LX1677	H179OSG	LX2055	H651YHT	LX1889	J376AWT	LX2103
G315EOG	LX1856	H180OSG	LX2056	H652YHT	LX1891	J377AWT	LX2104
G316DPA	LX1678	H181OSG	LX2057	H653YHT	LX1892	J379BWU	LX2114
G316EOG	LX1890	H182OSG	LX2058	H654YHT	LX1893	J380BWU	LX2115
G317NNW	LX1599	H183OSG	LX2059	H655YHT	LX1894	J381BWU	LX2116
G319NNW	LX1600	H184OSG	LX2060	H656YHT	LX1895	J382BWU	LX2117
G321NNW	LX1602	H185OSG	LX2061	H657YHT	LX1896	J564URW	LX2050
G322NNW	LX1603	H186OSG	LX2062	H658YHT	LX1897	J628LHF	LX2065
G324NNW	LX1601	H187OSG	LX2063	H659YHT	LX1898	J629LHF	LX2068
G324NUM	LX1630	H188OSG	LX2064	H660YHT	LX1899	J630LHF	LX2139
G326NUM	LX1632	H251GEV	LX1758	H661YHT	LX1900	J651UHN	LX2105
G327NUM	LX1633	H252GEV	LX1759	H662YHT	LX1901	J652UHN	LX2106
G329NUM	LX1598	H253GEV	LX1777	H733HWK	LX2003	J653UHN	LX2107
G330NUM	LX1769	H253PAJ	LX2091	H755WWW	LX2074	J654UHN	LX2108
G330NUM	LX1770	H254GEV	LX1796	H756WWW	LX2075	J655UHN	LX2109
G332NUM	LX1634	H254PAJ	LX2092	H757WWW	LX2076	J656UHN	LX2111
G381MWU	LX1270	H255GEV	LX1797	H813EKJ	LX1690	J657UHN	LX2112
G382MWU	LX1271	H256GEV	LX1798	H814EKJ	LX1691	J658UHN	LX2113
G383MWU	LX1272	H256YLG	LX2004	H815EKJ	LX1692	J724KBC	LX2121
G384MWU	LX1273	H257GEV	LX1799	H816EKJ	LX1693	J759DAU	LX2119
G472PGE	LX1307	H258GEV	LX1868	H851GRE	LX1857	J760DAU	LX2120
G473DHF	LX1326	H334TYG	LX2010	H852GRE	LX1858	J901UKV	LX2118
G473PGE	LX1316	H335TYG	LX2011	H853GRE	LX1859	J916WVC	LX2123
G474DHF	LX1327	H336TYG	LX2012	H854GRE	LX1860	J922MKC	LX2051
G508EAJ	LX1560	H337TYG	LX2013	H855GRE	LX1861	J923MKC	LX2052
G509EAJ	LX1561	H338UWT	LX2014	H856GRE	LX1862	J923MKC	LX2071
G510EAJ	LX1562	H339UWT	LX2015	H857GRE	LX1863	J924MKC	LX2069
G511EAJ	LX1563	H341UWT	LX2017	H858GRE	LX1864	J925MKC	LX2141
G512EAJ	LX1559	H342UWT	LX2018	H859GRE	LX1865	J926MKC	LX2070
G542CAC	LX1398	H343UWT	LX2016	H860GRE	LX1866	J927MKC	LX2073
G611GEF	LX1413	H343UWX	LX2028	H861GRE	LX1867	J928MKC	LX2072
G612GEF	LX1414	H344UWX	LX2029	HXI3006	B60.01	J929MKC	LX2140
G613GEF	LX1415	H345UWX	LX2030	HXI3007	LX1002	K10BMB	LX2136
G614GEF	LX1416	H346UWX	LX2031	HXI3008	LX1003	K11BMB	LX2137
G615GEF	LX1417	H347UWX	LX2032	HXI3009	LX1005	K12BMB	LX2138
G616GEF	LX1418	H355WWX	LX2077	HXI3010	LX1006	K27EWC	LX2048
G617GEF	LX1419	H356WWX	LX2078	HXI3011	LX1007	K101JMV	LX2133
G618GEF	LX1420	H357WWX	LX2079	HXI3012	LX1008	K622YVN	LX2122
G619GEF	LX1421	H358WWY	LX2080	J7JFS	LX2043	K623YVN	LX2124
G620GEF	LX1422	H359WWY	LX2081	J14JFS	LX2044	K624YVN	LX2125
G745PNN	LX1645	H393WWY	LX2083	J41GGB	LX2021	K625YVN	LX2126
G746PNN	LX1646	H407ERO	LX1869	J42GGB	LX2022	K626YVN	LX2127
G747PNN	LX1647	H408ERO	LX1870	J43GGB	LX2024	K627YVN	LX2128
G748PNN	LX1648	H408YMA	LX1318	J45GGB	LX2047	K628YVN	LX2129
G749PNN	LX1649	H409ERO	LX1871	J249KWM	LX2042	K629YVN	LX2130
G803EKA	LX1400	H410ERO	LX1872	J250KWM	LX1195	K630YVN	LX2131
G936VRY	LX1277	H460WWY	LX2082	J251KWM	LX2066	K801CAN	LX2134
G992VWV	LX1269	H542FWM	LX1424	J261UDW	LX2023	K802CAN	LX2135
G993VWV	LX1278	H543FWN	LX2005	J262UDW	LX2025	K852MTJ	LX2067
G994VWV	LX1279	H544FWN	LX2007	J263UDW	LX2026	K853MTJ	LX2142
G995VWV	LX1280	H610YTC	LX1902	J264UDW	LX2033	MO8500	LX1848
G996VWV	LX1281	H611YTC	LX1903	J265UDW	LX2034	SBS3572Y	LX1136
H23YBV	LX1394	H612YTC	LX1904	J266UDW	LX2035	U/reg	B60.04
H24YBV	LX1395	H613YTC	LX1905	J267UDW	LX2036	U/reg	B60.06
H26YBV	LX1396	H614YTC	LX1906	J268UDW	LX2038	U/reg	LX1001
H27YBV	LX1397	H615YTC	LX1907	J269UDW	LX2039	U/reg	LX1015
H28MJN	LX1694	H616YTC	LX1908	J270UDW	LX2045	U/reg	LX1153
H28YBV	LX1399	H633YHT	LX1842	J271UDW	LX2046	U/reg	LX1211
H29MJN	LX1695	H634YHT	LX1873	J295TWK	LX2049	U/reg	LX2001
H29YBV	LX1401			J362YWX	LX2084	U/reg	LX2037
H31PAJ	LX2088						

British Bus Publishing

BUS HANDBOOKS

Also available!

The 1996 First Bus Handbook - £9.95
The 1995 Stagecoach Bus Handbook - £9.95
The 1996 Stagecoach Bus Handbook - £9.95
The North East Bus Handbook - £9.95
The Yorkshire Bus Handbook - £9.95
The Scottish Bus Handbook - £9.95
The Welsh Bus Handbook - £9.95
The East Midlands Bus Handbook - £8.95
The North Midlands Bus Handbook - £8.95
The South Midlands Bus Handbook - £8.95
The Model Bus Handbook - £9.95
The Fire Brigade Handbook - £8.95

Coming Soon
**Cumbria, Lancashire and Manchester
Bus Handbook - £9.95
Merseyside and Cheshire Bus Handbook - £9.95**

Get the best!
Buy today from your local bookseller,
or order direct from:

**British Bus Publishing
The Vyne, 16 St Margaret's Drive, Wellington
Telford, Shropshire TF1 3PH**
Fax and Credit Card orders: 01952 255669